THE COSMOS OF KHNUMHOTEP II

STUDIES IN EGYPTOLOGY

EDITED BY GEOFFREY THORNDIKE MARTIN

EDWARDS PROFESSOR OF EGYPTOLOGY EMERITUS,
UNIVERSITY COLLEGE LONDON

THE EGYPTIAN TEMPLE
PATRICIA SPENCER

THE ADMINISTRATION OF EGYPT IN THE OLD KINGDOM
NIGEL STRUDWICK

CORPUS OF RELIEFS OF THE NEW KINGDOM FROM THE MEMPHITE NECROPOLIS AND LOWER EGYPT, VOLUME 1
GEOFFREY THORNDIKE MARTIN

PROBLEMS AND PRIORITIES IN EGYPTIAN ARCHAEOLOGY
EDITED BY JAN ASSMANN, GÜNTER BURKARD AND VIVIAN DAVIES

LOST TOMBS
LISE MANNICHE

DECORATION IN EGYPTIAN TOMBS OF THE OLD KINGDOM
YVONNE HARPUR

UNTERSUCHUNGEN ZU DEN TOTENBUCHPAPYRI DER 18. DYNASTIE
IRMTRAUT MUNRO

THE MONUMENTS OF SENENMUT
PETER F. DORMAN

THE FORT-CEMETERY AT HIERAKONPOLIS
BARBARA ADAMS

THE DUTIES OF THE VIZIER
G. P. F. VAN DEN BOORN

A GLOSSARY OF ANCIENT EGYPTIAN NAUTICAL TITLES AND TERMS
DILWYN JONES

LAND TENURE IN THE RAMESSIDE PERIOD
SALLY L.D. KATARY

VALLEY OF THE KINGS
C.N. REEVES

THE COBRA GODDESS OF ANCIENT EGYPT
SALLY B. JOHNSON

A BIBLIOGRAPHY OF THE AMARNA PERIOD AND ITS AFTERMATH
GEOFFREY THORNDIKE MARTIN

THE PRIVATE CHAPEL IN ANCIENT EGYPT
ANN H. BOMANN

AKHENATEN'S SED-FESTIVAL AT KARNAK
JOCELYN GOHARY

AFTER TUT'ANKHAMUN
EDITED BY C.N. REEVES

THE BOUNDARY STELAE OF AKHENATEN
WILLIAM J. MURNANE AND CHARLES C. VAN SICLEN III

THE CANOPIC EQUIPMENT OF THE KINGS OF EGYPT
AIDAN DODSON

LIVING IN THE PAST: STUDIES IN ARCHAISM OF THE EGYPTIAN TWENTY-SIXTH DYNASTY
PETER DER MANUELIAN

EGYPTIAN SOLAR RELIGION IN THE NEW KINGDOM
JAN ASSMANN

WINE AND WINE OFFERING IN THE RELIGION OF ANCIENT EGYPT
MU-CHOU POO

ASKUT IN NUBIA
STUART TYSON SMITH

THE NEW KINGDOM ROYAL CITY
PETER LACOVARA

CHIEF OF SEERS
ELIZABETH GORING, NICHOLAS REEVES AND JOHN RUFFLE

THE COSMOS OF KHNUMHOTEP II
AT BENI HASAN

JANICE KAMRIN

KEGAN PAUL INTERNATIONAL
LONDON AND NEW YORK

to my three grandmothers

First published in 1999 by
Kegan Paul International
UK: P.O. Box 256, London WC1B 3SW, England
Tel: (0171) 580 5511 Fax: (0171) 436 0899
E-mail: books@keganpau.demon.co.uk
Internet: http://www.demon.co.uk/keganpaul/
USA: 562 West 113th Street, New York, NY, 10025, USA
Tel: (212) 666 1000 Fax: (212) 316 3100

Distributed by
John Wiley & Sons Ltd
Southern Cross Trading Estate
1 Oldlands Way, Bognor Regis
West Sussex, PO22 9SA, England
Tel: (01243) 779 777 Fax: (01243) 820 250

Columbia University Press
562 West 113th Street
New York, NY 10025. USA
Tel: (212) 666 1000 Fax: (212) 316 3100

Printed in Great Britain

British Library Cataloguing in Publication Data
Kamrin, Janice
The cosmos of Khnumhotep II at Beni Hasan
1. Khnumhotep, II - Tombs 2. Bani Hasan Site (Egypt) 3. Egypt - Antiquities
I. Title
932
ISBN 0-7103-0574-5

Library of Congress Cataloging-in-Publication Data
Kamrin, Janice, 1966-
The cosmos of Khnumhotep II at Beni Hasan / Janice Kamrin.
-- (Studies in Egyptology)
Includes bibliographical references (p.) and index.
ISBN 0-7103-0574-5
1. Banī Hasan Site (Egypt) 2. Khnumhotep II--Tomb. 3. Cosmology, Egyptian.
I. Title. II. Series.
DT73.B4K36 1998
932--dc21 98-42178
CIP

TABLE OF CONTENTS

LIST OF FIGURES

LIST OF TABLES

LIST OF PLATES

ACKNOWLEDGEMENTS

Writing a book is a long and sometimes agonizing process, and cannot be done alone. I am indebted to many people for their help, guidance, and moral support during the years it has taken me to produce this volume.

I wish to thank, first and foremost, Dr. David O'Connor, whose contributions to this work are immeasurable. He gave unstintingly of his time and expertise, and was generous both with his knowledge and with his ideas. His enthusiasm for Egyptology is contagious, and I always left our meetings revitalized for the task at hand. I would also like to thank Holly Pittman, Donald Spanel, and David Silverman, all of whom have contributed substantively to this work.

My husband, John Immerwahr, has been enormously supportive of this work, and has helped in many ways, from moral, practical, and financial support (including wonderful meals) to acting as unofficial reader, checking my writing from time to time to make sure it still made sense,. He and our boys, Daniel and Adam, have been wonderful.

Janet Richards has traveled with me every step of the way, from the beginning of my coursework. She has always been there for me when I needed either personal or professional advice.

My graduate studies in Egyptology would never have begun without the inspiration of Brunilde Ridgway, who has served in many ways as both mentor and friend to me, and the guidance of Machteld Mellink, Richard Ellis, and Gloria Pinney. My years at Bryn Mawr were exciting and challenging, and both the knowledge and the scholarship which were instilled in me there have stood me in good stead in my work and in my life.

Bonnie Crossfield, without whom the Museum would cease to function, has also been a great help to me. She has always made me feel welcome, and dealt with a multitude of problems for me.

Jean Adelman and Anita Fahringer have always been helpful and supportive, and have kept the library running smoothly throughout my years at Penn. In the last year or so, Mark Lewis has also added a cheerful presence. Peggy Guinan and Diane Bergin kept the paperwork flowing and have lent me moral support and guided me through all the red tape.

My years at Penn were made enjoyable by my fellow students, especially Stacie Olson, Melissa Robinson, Joe Wegner, Steven Harvey, Matt Adams, Brian Muhs, and Allison Webb. Zahi Hawass brightened my early years here, and provided me with much inspiration. He has also facilitated the time I have spent in Egypt, for which I am very grateful.

I would also like to thank my family, Mom, Dad, Phyllis, and Audrey, for the support they have given me over the years. My parents instilled in me a love for education in general, and introduced me to the world of archaeology at an early age. I would like to thank Dorothy and Shelby Rooks for their continual love and interest, and Dr. Philip Cho for helping to keep me sane and happy during the period during which most of this book was written. And last but not least, I would like to thank Dan Yankelovich for the moral and practical support he has given me during the final stages of this project.

CHAPTER I : INTRODUCTION

THESIS STATEMENT

High in a cliff at the remote site of Beni Hasan in Middle Egypt, thirty-nine ancient tombs line a narrow ledge above the Nile River. These tombs were cut into the rock face, with pillars of living rock often left standing inside the echoing chambers. Each tomb consists of a one or two-chambered chapel, into the floor of which were cut two or more deep shafts which lead to subterranean burial chambers. A number of the chapels are richly painted, their walls covered with lively scenes depicting the ancient nobles, their families, and the workers on their estates engaged in various tasks, accompanied by brightly colored depictions of birds, fish, and animals of the desert. I have chosen one of the largest and most richly decorated of these chapels, the tomb chapel of Khnumhotep II, to study in depth; this monument is securely dated, well-preserved, and reasonably comprehensively published.

The tomb of Khnumhotep II dates to the early 12th Dynasty, primarily to the reign of Amenemhēt II. The owner was a high official of the ancient administrative area in which Beni Hasan was located, the Oryx (XVIth Upper Egyptian) nome or province. His primary title was Overseer of the Eastern Desert, a title which he held from Year 19 of Amenemhēt II (c. 1910 B.C.) until at least Year 6 of Senwosret II (c. 1891 B.C.). This monument is the latest of the large Beni Hasan tombs, and represents the culmination of the series.

The detailed analysis of this complex tomb necessarily comprises most of this volume. However, the results of this study may well apply to other tombs. As an evaluation of this possibility, and as a check on the general validity of the conclusions, a second tomb from Beni Hasan will also be briefly described and discussed. The tomb chosen for comparative purposes is Tomb 2, which belonged to a man named Ameni, Nomarch (governor) of the Oryx

Nome from Year 18 to Year 43 of Senwosret I (c. 1973-1948 B.C.). The extensive decoration of this tomb is quite well preserved and has been published in its entirety; it is discussed in Chapter V.

The tomb chapel of Khnumhotep II is entered from a natural ledge along the face of the cliff. This ledge was widened through the process of cutting the tombs, so that the chapel is fronted by a columned portico and a small courtyard. This courtyard was once surrounded by mud-brick walls of now uncertain dimensions. The rock face around the entrance to the tomb chapel was smoothed and flattened, and the doorway to the interior was set into the center of this façade. The chapel is a large rectilinear chamber cut horizontally into the cliff; six columns (now mostly destroyed) were left standing in the living rock. The ceiling consists of three east-west barrel-shaped vaults cut into the rock. At the back of the main chamber, along the main east-west axis of the tomb, is an additional small chamber, the shrine. Two large shafts leading to complexes of burial chambers far below were cut into the floor of the main chamber.

The visitor to this tomb is immediately struck by the lavishness of the painted decoration, which covers the walls and the ceiling with rich color. The effect is that of a complex tapestry, with figures and text woven together into elegant patterns. But what does this decoration represent, and what might the ancient Egyptians have intended as its function? The first stage in understanding this decoration is reached by identifying the activities depicted and translating the texts. Once this groundwork has been laid, other questions can be asked: Why was the tomb decorated at all? What roles do these brightly colored paintings play? How has the decoration been chosen and organized? How does the decoration help to express the function of the tomb chapel within the larger Egyptian world-view? These questions form the core of this research.

The function of the decoration of elite, or high-status, tomb chapels of the Egyptian Old and Middle Kingdoms has often been thought by scholars to have provided for the physical needs of the deceased in eternity, and to have guaranteed the perpetual enjoyment of things or activities the noble had enjoyed in life. The tomb is thus seen as a "house," in which the noble would live for eternity. This explanation, however, does not account for much of the content of the decoration of such structures. Alternative explanations of the decoration, which would not necessarily exclude the application of traditional views, must be sought. Much work has been done, some very recently, which demonstrates that various cult structures, specifically the gods' temple, the royal mortuary temple, and the royal ka chapel, can be seen as cosmograms, microcosmic and schematized representations of the Egyptian cosmos. The complex and meaningful architectural and decorative programs carried out within these structures were chosen not for aesthetic reasons, but in order to relate the structure effectively to the powers and operations of the cosmos.

Study of the elite tomb chapel can benefit greatly from this type of approach. The goal of this work is therefore to suggest ways in which the architectural form and the decoration of the elite tomb chapel, at least of the 12th Dynasty, may have reflected aspects of the Egyptian conception of the cosmos and the relationship of the tomb owner to it.

In order to test and explore the implications of the hypothesis that the Egyptian elite tomb can also be a cosmogram, this work offers a detailed study of the decoration of Khnumhotep II's tomb chapel, including scenes, ornamental features, and texts. Various levels of meaning contained within the components of this decoration are discussed. These levels range from the actual, where a figure or activity can be identified as corresponding to the realities of Egyptian life on earth, to the symbolic, where a figure or activity might represent, for example, an element from the realm of Egyptian mythology. Finally, the ways in which these components may have been organized and interrelated to express one or more aspects of the cosmos or cosmoi in which the deceased noble functioned are addressed.

This work has led me to the conclusion that the tomb chapel of Khnumhotep II, and to some extent also the tomb chapel of Ameni, can be interpreted as a miniature model of the cosmos which integrates the worlds of the living and the dead, in which the events depicted and rites carried out help guarantee the eternal survival of the noble and the proper functioning of the Egyptian world. In fact, in this tomb, there are three interrelated cosmoi represented: the personal cosmos of Khnumhotep II, the royal cosmos, and the larger Cosmos. In a way, the tomb chapel can be thought of not only as a "house," but also as a private "mortuary temple" for the deceased noble, where he would dwell for eternity as a being with divine characteristics, effective outside the earthly realm. Within this chapel, he would receive offerings, and in return he would act to help maintain the proper order of the cosmos.

THEORETICAL FRAMEWORK: THE ANCIENT EGYPTIAN CONCEPTION OF THE COSMOS

An understanding of the Egyptian view of the cosmos during the Middle Kingdom is vital for this analysis. This view is inevitably that of the elite, documented in texts and other sources. Whether it is also the view of the lower classes is a subject for future research. Moreover, there was variation over time in the Egyptian conception of the cosmos, and the 12th Dynasty was a period of substantial change in many respects. In addition, in any single period, multiple contradictory views would seem to be held on features of this cosmos without seeming to conflict with one another.[1] Keeping these issues in mind, we can see that careful study of religious scenes and texts of the various

[1] Frankfort, Ancient Egyptian Religion, pp. 3ff.

pharaonic periods has allowed scholars to identify many aspects of the ancient Egyptian conception of the cosmos,[2] which included 1) cosmogony and cosmic process, 2) cosmography, 3) a hierarchy or continuum of beings, and 4) a repertoire of built structures, referred to here as the architectural environment, all elements of importance to this study.

COSMOGONY AND COSMIC PROCESS

Before the Creation, the world was thought to consist of the formless Nun, conceptualized as limitless darkness and water. At the time of Creation, the eternal and unchanging pattern of life and existence, of light and order, called in Egyptian *maat (m3ꜥt),* was brought into being from Nun through the actions of the creator god. Nun was therefore seen as the source of all life, "existence waiting to happen."[3] This Creation was thought to have a distinct beginning and end, and its unique occurrence was labeled the "first time" *(sp tpy).* However, the chaotic, formless waters of Nun surrounded the created universe and continually threatened its existence; Nun was therefore also seen as the negation of life and order.[4] Because of the continually threatening powers of Nun, called *isfet (isft), maat,* the order of the universe, had to be renewed perpetually by the daily re-enactment of the original Creation, played out primarily by the daily cycle of the sun and the yearly cycle of the flood.[5]

Within this created universe, two types of time appear to have functioned. The enduring, unchanging world, in which the cosmos has a finite beginning and an end, was linear, or *djet* (*ḏt*) time; this type of eternity could also be considered timelessness, and was linked with the god Osiris. Cyclical time, called *neheh (nḥḥ),* was expressed through the continual re-enactment of Creation, embodied by the daily cycle of the sun, the yearly cycle of the seasons, and the perpetual death and rebirth of the sun god.[6]

COSMOGRAPHY

Geographically, the created cosmos was divided into the earth, the sky, and the *Duat (dw3t).* Earth seems to have been visualized as a huge flat disk,[7] divided into environmentally differentiated regions. Its central feature was the Nile River, running South to North through its center. The river was flanked by cultivated land, on which the estates of the nobles, the palaces of the kings,

---

[2] Hornung, Idea into Image, pp. 39-40.

[3] Allen, Genesis in Egypt, p. 57.

[4] Allen, Genesis in Egypt, p. 57. This concept, at least as it is expressed by Egyptologists, seems almost modern: the ancient concept of chaos is similar to the modern concept of entropy, whose natural tendencies are toward disorder in the sense of undifferentiation.

[5] Assmann, "Schöpfung," LÄ VI, col. 685.

[6] Allen, Genesis in Egypt, pp. 31ff.; Wente, "Funerary Beliefs," p. 22. Cf. however, Hornung, Idea into Image, pp. 64ff.

[7] Schäfer, Principles, pp. 235-37.

the temples of the gods, and the homes of ordinary people were generally located. It was here in the floodplain that the agriculture which supported Egyptian life was carried out, and where activities such as manufacture, food preparation, and food storage took place. The floodplain was convex in shape, so that the lowest-lying areas were generally located at its outer edges. The low-lying areas at the margins of the valley often would have retained water year-round, and would generally have had a swampy, marshy character.[8]

Figure I.1: Schematic View of the Nile Valley Landscape

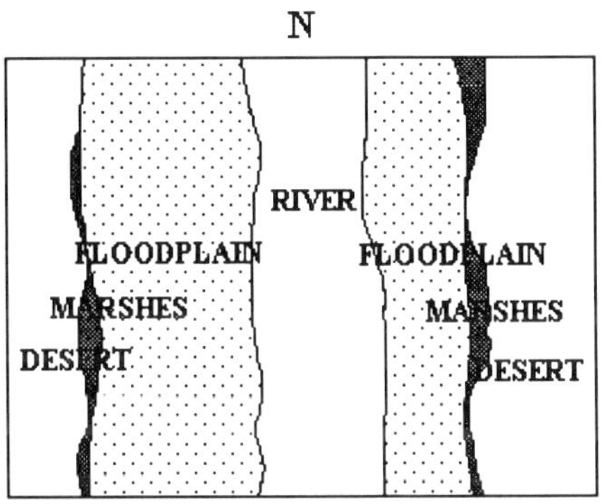

Fringing the floodplain was the low desert, which may have been somewhat less arid in the early Middle Kingdom than it is now.[9] This region was populated by animals such as wild cattle, gazelles, oryxes, wild goats, jackals, and lions. The desert, because of its position outside the floodplain and its inhospitable nature, was considered chaotic, an area where the forces of disorder that surrounded the created world could and did intrude. Also outside the land of Egypt, but still on earth, was the rest of the physical world, inhabited by foreigners who were, like the desert animals, considered to be representative of the intrusive and disorderly forces of chaos.

The cycle of seasons was played out within the realm of the earth. There were three main seasons in the Egyptian calendar, each of which was divided into four months of thirty days; an additional five days were added to the end of the year. *Akhet (ꜣḫt)*, or "Inundation," which began the year, was heralded by the appearance of the Dog-star, Sothis, in the sky. The flood began in mid July, and the floodplain remained under water throughout late August and September. When the flood receded in the late fall, boundaries were re-established and crops were planted (beginning in November). This was the season of *Peret (prt)*, or "Going Forth" (probably referring to the emergence of

---

[8] Butzer, Hydraulic Civilization, pp. 16-18.
[9] Butzer, Hydraulic Civilization, pp. 26-27.

the fields from under the flood waters). The third season was *Shemu, (šmw)*, "Harvest," which ran from March to July. The early part of this season was the time of harvest; by its end, the weather was very hot, the fields lay fallow, and the river fell to its lowest levels.[10]

Figure I.2: The Seasons of the Egyptian Year

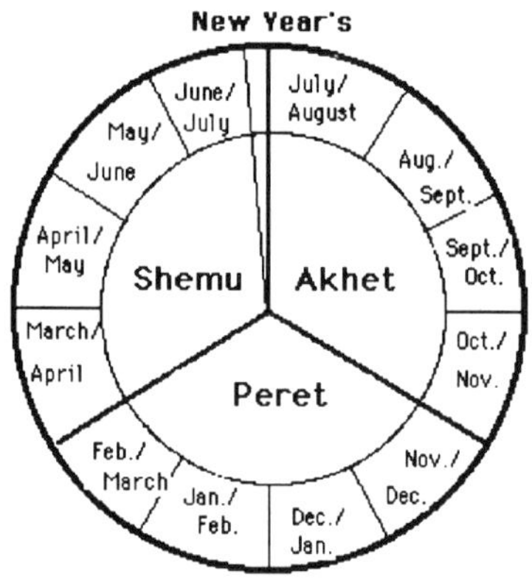

The sky, located above the earth, was conceived of in the form of the *pt* sign. It rested on four pillars and was separated from the earth (Geb) by the air, personified by the god Shu.[11] The sky, like the earth, was thought to contain a river, or to be bounded by water, on which the sun sailed during the day, moving from East to West, and where the stars traveled at night.[12] The sun, anthropomorphized as a solar deity, was imagined to rise in the eastern desert and set in the western desert. On his journey, the sun god would pass through the eastern and western horizons, which were considered transitional areas between the earthly realm and the Hereafter, or *Duat*.[13] Two celestial fields, the Field of Reeds and the Field of Offerings, were thought to be located at the two horizons.[14]

The region outside direct human experience, where the sun went at night and human beings went after death, was conceptualized as the *Duat*. Egyptian texts are ambiguous as to the exact location of this area, but it seems to have existed simultaneously both below the earth, as the netherworld, and above it, in the night sky, which took the form of the goddess Nut.[15] It is

---

10 Gardiner, Grammar, p. 203; von Beckerath, "Kalender," LÄ III, cols. 296-99; Parker, Calendars.

11 Schäfer, Principles, pp. 235-37.

12 Allen, Genesis in Egypt, p. 56.

13 Westendorf, "Weltbild," LÄ VI, col. 1211.

14 Lesko, "Cosmogonies," p. 120; Allen, "Cosmology of the Pyramid Texts," p. 6.

15 Lesko, "Cosmogonies," pp. 118-19.

described in some detail in funerary texts of the Middle and New Kingdoms as a dangerous region, full of demons and traps for those who had to pass through it. At the same time, it was full of the power of regeneration: the sun god and deceased humans were reborn after passing through the *Duat*.[16]

Figure I.3: "Geography" of the Egyptian Cosmos

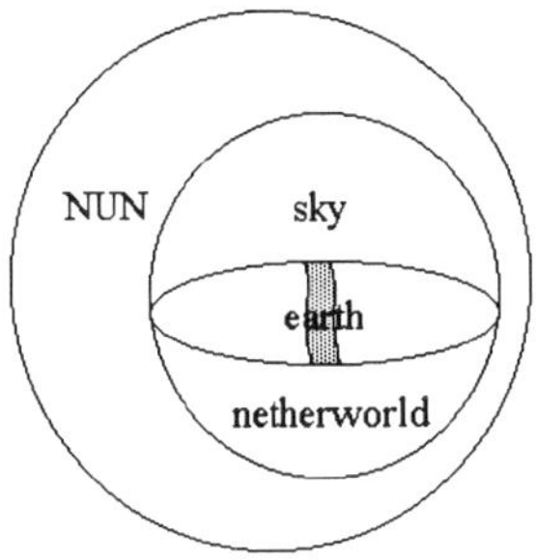

HIERARCHY OF BEINGS

Within this created cosmos was a hierarchy of beings: the creator god, the sun god, the gods, the king, the nobility, and the common people (Egyptians). The creator god was the foremost of the beings who dwelled within the Egyptian universe. This god was thought to have existed before the Creation, and was responsible for the differentiation of the cosmos from the primordial Nun and the existence of all the other beings in the universe. The mechanism for the Creation and the identity of the creator god could vary, but the importance of his role in the universe always transcends that of all other beings.[17]

The first entities brought into being by the creator god were the gods, chief among whom was the sun god. This god, who was most often identified as a form of Re, was responsible for the daily re-enactment of the Creation as expressed by the solar cycle, and was therefore a frequently seen embodiment of the creator god. The numerous gods in the Egyptian pantheon represented particular aspects of the universe, and played various roles which, taken in their totality, ensured the effective action of the cosmos. Through their existence and their actions, the gods maintained the proper cosmic order, *maat*, and held the forces of chaos at bay.[18]

There are many important Egyptian gods, only a few of whom need be discussed in this study. In addition to the sun god, who ruled in the celestial

---

16 Allen, Genesis in Egypt, p. 57; Allen, "Funerary Texts," p. 41.

17 Allen, Genesis in Egypt, pp. 48ff.

18 See Hornung, Conceptions of God, pp. 203ff.

realm, mention should be made of the god Osiris, who functioned primarily as the ruler of the netherworld. In Egyptian mythology, Osiris was the prototypic god who experienced death and resurrection: he is killed by his brother and resurrected by his sister/wife Isis.[19] In the New Kingdom at least, the sun god is identified with Osiris during his nightly journey through the netherworld, thus enabling him to become reborn.[20]

At the time of the Creation, the creator god also brought into being the institution of kingship, of dominion over the earth. According to both king-lists and mythological texts, the kingship was held by a succession of gods before it passed to Osiris, who in turn became, after his death and resurrection, ruler of the netherworld.[21] After a series of mythological conflicts between Seth and Horus, the son of Osiris and the goddess Isis, the earthly kingship was granted to Horus, who became the prototypic embodiment of the king.[22]

The living king, who dwelt on earth, was ascribed different characteristics according to the particular contexts in which he was seen. His primary roles are expressed in one stanza of a royal text dating probably to the Middle Kingdom:

iw rdi.n Rꜥw niswt N
tp tꜣ n ꜥnḫw
r nḥḥ ḥnꜥ ḏt
ḥr wḏꜥ rmṯw ḥr sḥtp nṯrw
ḥr sḫpr mꜣꜥt ḥr sḥtm isft
iw di.f ḥtpwt n nṯrw
prt-ḫrw n ꜣḫw

Re has installed king N
upon the land of the living
for ever and eternity
judging men, satisfying gods,
realizing *maat*, annihilating *isfet*.
He gives offerings to the gods
and mortuary offerings to the dead.[23]

---

19 Frankfort, Kingship and the Gods, Chapter 15.

20 Hornung, Conceptions of God, pp. 155-56.

21 Frankfort, Kingship and the Gods, Chapter 15.

22 Frankfort, Kingship and the Gods, pp. 197ff.; Griffiths, The Conflict of Horus and Seth; te Velde, "Horus und Seth," LÄ III, cols. 25-27. Horus is himself a very complex deity who, in addition to and as part of his royal role, was a very important sky god (see also Schenkel, "Horus," LÄ III, cols. 14-23; Hornung, Conceptions of God, pp. 143ff.).

23 Transliteration and translation, as well as dating, are taken from Assmann, "State and Religion," p. 58.

According to this text, therefore, the living king meted out justice to his subjects, and was ultimately responsible for their well-being. In gods' temples, it is the king who acts as chief priest, giving offerings and thereby satisfying the deities. He is also theoretically responsible for the mortuary cults of his subjects, a responsibility expressed by the *hetep-di-nesu (ḥtp-di-nsw)* formula which features prominently in private tombs.[24] In cultic contexts, such as god's temples and royal mortuary temples, the king is seen as the guarantor of the cosmic order, maintaining *maat* and symbolically repelling *isfet* by actions such as dominating foreigners or hunting desert animals.[25]

The divinity of the king during his lifetime is a complex issue. He was not divine from birth, but acquired divine attributes through rituals celebrated at his accession to the throne, at which time he was seen as the embodiment of the prototypic king.[26] Beginning at least in the 4th Dynasty, the king is also identified explicitly as the son of the sun god through the title *s3 Rꜥ*, "son of Re."[27] This identification becomes even more explicit in the New Kingdom, when the divine birth of the pharaoh was given great prominence. Although the king was identified with various gods, and was given epithets such as *netjer nefer (nṯr nfr)*, "the good god," he was not considered a deity equal in capacity to the gods. Even during his tenure as king, he was subordinate to the gods, and was granted his power to rule in return for his actions as chief priest to them. Because of his dual nature as both human and divine, the king served as intermediary between the gods and humankind; his existence was essential to the proper cosmic order.[28]

In contrast to his role as the living king, the deceased king was more securely assured of divinity, although he was still subordinate to the gods under most circumstances. As expressed in royal funerary sources such as the Pyramid Texts, the king was closely identified after death with various gods, in particular Re and Osiris. As Osiris, he was linked with linear eternity *(djet)*, while as Re, he helped to ensure the daily cycle of regeneration *(neheh)*.[29] The king therefore continued after his death to function as a guarantor of the cosmic order.

Ranked below the king in the social hierarchy were the high nobility, represented by court officials and provincial rulers such as Khnumhotep II of

---

24 For discussion of this formula, see Gardiner, Grammar, pp. 170ff.; Barta, Opferformel; Lapp, Opferformel; and Bennett, "Growth of the *Ḥtp-di-nsw* formula." The standard translation of this phrase reads: "An offering which the king gives...to N."

25 Blumenthal, "Königsideologie," LÄ III, cols. 528-29.

26 Frankfort, Kingship and the Gods; Silverman, "Divinity and Deities;" Silverman, "The Nature of Egyptian Kingship."

27 Cf. von Beckerath, Handbuch der ägyptischen Königsnamen, p. 328.

28 Cf. Hornung, Conceptions of God, pp. 141-142; Silverman, "Divinity and Deities," pp. 58ff.; Silverman, " The Nature of Egyptian Kingship."

29 Silverman, "Divinity and Deities," p. 72.

Beni Hasan. As expressed in their titles, these elite could share in some royal functions. For example, Khnumhotep II held a number of priestly titles connected with temples of Horus, of Anubis, and of Pakht.[30] He thus represented the king, who was ultimately the chief priest of these cults. Nobles were also responsible for administering justice within their jurisdictions, another attribute which belonged ultimately to the kingship. The living noble could thus be seen as embodying the king in certain spheres, as the king could be identified with various gods in specific contexts.

Like the deceased king, the noble acquired further attributes of divinity after his death and proper burial. The relative divinity of the deceased noble appears to have undergone considerable change over the course of Egyptian history, and is of considerable importance to this study. The Coffin Texts, which probably developed from the Pyramid Texts of the Old Kingdom, may have been used by high status people as early as the 6th Dynasty, and were relatively widespread by the Middle Kingdom.[31] These funerary texts, examples of which have been found on Middle Kingdom coffins from Beni Hasan, provided for the divine transformation of the deceased noble and his or her identification with Osiris, thereby bestowing on them prerogatives that had originally been royal.[32] In the New Kingdom, with the so-called "democratization" of the Afterlife, divine attributes became even more widely available to non-royal and even non-elite individuals.[33]

The decoration of private tomb chapels at the end of the Old Kingdom and in the Middle Kingdom also displays some usurpation of royal and divine attributes. Scenes which depict the noble hunting in the marshes, for example, are based on royal mortuary prototypes,[34] and represent a distinct shift in "decorum," a term which can be defined as the allowable representation of access to the divine.[35] A change in permissible access to the divine from the royal to the private realms is also expressed in tomb biographies of the Old and Middle Kingdoms.[36] The shift in decorum which occurs during the 12th Dynasty is well represented by the tomb chapel of Khnumhotep II at Beni Hasan, and will be discussed further below in the context of the discussion of this monument.

---

[30] Newberry, BH I, Pl. 24, northern architrave, line b, and shrine, left and right jambs.

[31] See Silverman, "Divinity and Deities," p. 72, note 86.

[32] Frankfort, Ancient Egyptian Religion, pp. 103ff; Frankfort, Kingship and the Gods, p. 90.

[33] Sorenson, "Divine Access," pp. 117ff.; and Silverman, "Divinity and Deities," pp. 72-73.

[34] See below, microscenes E1 and E3.

[35] Baines, "Society, Morality, and Religious Practice," pp. 125-27; Baines, Fecundity Figures, pp. 277ff; and Sorensen, "Divine Access," pp. 109-117.

[36] Altenmüller, "Lebenszeit und Unsterblickheit," pp. 78ff.

THE ARCHITECTURAL ENVIRONMENT

Included within the Egyptian universe, specifically on the earth but in some cases providing links to other areas of the cosmos, were a number of built structures. The living king dwelt in a palace, while the living noble lived in a house. The deceased king was buried in a royal tomb, and worshipped in a royal mortuary temple. In the Old and Middle Kingdoms, these temples were part of the burial complexes of their respective kings; in the New Kingdom, they were separate structures.[37] The deceased noble also had a tomb, and an associated place of worship, a tomb chapel. Gods were worshipped in gods' temples. Royal palaces could be associated with these gods' temples, as could both royal and private ka chapels, which were dedicated to deceased kings or nobles.[38]

Figure I.4: Built Structures in the Egyptian Cosmos

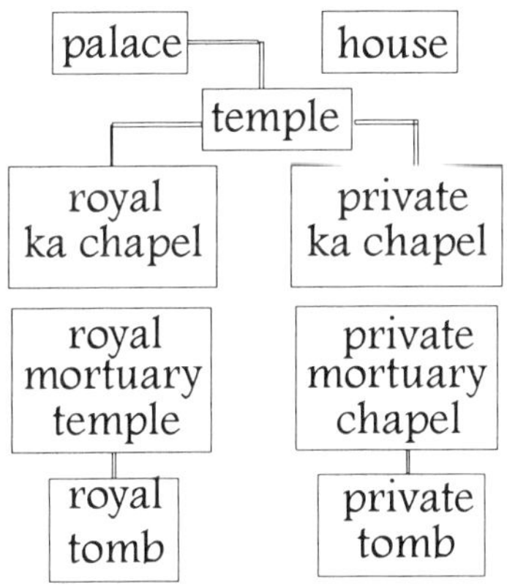

These various types of structures have a complex interrelationship. The temple was considered the dwelling of the god, as the house was the dwelling of the living noble and the palace the home of the living king. The tomb has also been seen as the house of the deceased. As noted by Gardiner in 1935: "...temple, tomb, and house of the living all bore a strong resemblance to one another, containing rooms where the owner lived, and others where his possessions were stored."[39]

Most comparisons of temple or tomb to house are drawn from the Middle Kingdom houses at Kahun and the New Kingdom estates at Amarna, as well as to some extent from royal palaces.[40] The basic model is as follows. The palace or elite house consists of three main areas: a large open court with a

---

[37] Spencer, Death in Ancient Egypt, p. 63.
[38] O'Connor, "City and Palace," and "The Status of Early Egyptian Temples."
[39] Gardiner, Attitude, p. 10.
[40] Steindorff, "Haus und Tempel," p. 108; Kemp, Anatomy, pp. 151f.

pillared portico at the back; a broad columned hall; and a narrow deep area, which probably represents the private eating and sleeping chambers of the house owner and his family. Many other rooms with various functions, such as storerooms, are arranged behind and around this core.[41]

Figure I.5: Plan of a Typical New Kingdom God's Temple[42]

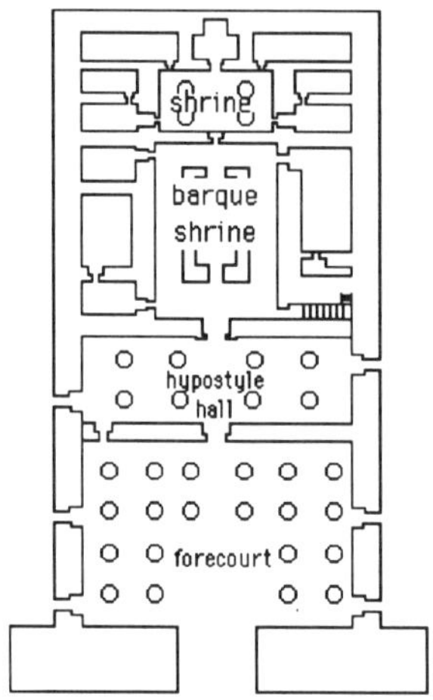

As demonstrated first by Steindorff in 1896[43] and reiterated and refined in later publications,[44] the architectural layout of both gods' temples and royal mortuary temples of the New Kingdom make overt reference to the plan of a palace or elite house. The house and temple both contain the same basic elements: courtyards, pillared halls, and inner apartments, which include a "dining room," storerooms, and private chambers for the king, noble or god and his or her family. This correlation can be seen, for example, in the Khonsu temple at Karnak. Just behind the pylon which defines the entrance to the temple lies an open court, bordered by pillared porticos. Beyond the open court is a broad pillared hall which spans the breadth of the building, and then a narrow, deep chamber, which forms the inner sanctuary where the god receives offerings. Behind and around the sanctuary are other rooms with various functions, such as magazines.

The similarity between the major elements of the two types of structures is striking, although the layouts are not particularly comparable. The functions

---

41 Kemp, Anatomy, pp. 151f.

42 Temple of Khonsu at Karnak (After Porter and Moss, Topographical Bibliography II, Pl. 21).

43 Steindorff, "Haus und Tempel."

44 Cf. Arnold, Wandrelief und Raumfunktion.

of the corresponding areas in house and temple are of primary significance. The concordance between house and temple is also expressed in an Egyptian word for temple: "*ḥwt-nṯr*," "house/estate of the god."[45]

The explicit conception of the private tomb as a house dates back into the Early Dynastic period, as seen most clearly in several tombs at Saqqara which contain subterranean chambers arranged to resemble the rooms of a palace or elite house, including bedrooms, lavatories, and storage areas.[46] The superstructures of many Early Dynastic tombs, both royal and elite, are also thought to be modeled on the palace, as expressed through the "palace façade" niching which ornaments the outside.[47]

The architectural correspondence of the tomb with a contemporary house apparently receded into the background for much of the Old Kingdom, a time of considerable architectural variation,[48] but revived toward the end of the 5th Dynasty and became increasingly evident during the 6th Dynasty.[49] The architectural correlation is most evident in rock-cut and free-standing tombs of a certain type (fronted by porticos); the layout represented by the tomb of Khnumhotep II at Beni Hasan can clearly be shown to develop from both types.[50] One of the earliest examples of this type is represented by the 5th Dynasty tomb of Nysedjerkai at Giza. This tomb is entered through a wooden door which leads into an open court, then to a pillared portico whose roof has gutters for rain water. From here, one enters a broad pillared hall, corresponding to the entry hall for visitors seen in the later houses from Amarna. In this hall, there are two false doors, which correspond to the doors in a house which lead to the private chambers, but here just imply such rooms. Behind these false doors there is simply a large chamber in the center of which are two burial shafts.[51]

Since both temples and tombs can usefully be compared to the model of the house, it seems reasonable to hypothesize that there are correlations between tombs and temples. The architectural resemblance between rock-cut tombs of the Middle Kingdom, such as the tomb of Khnumhotep II at Beni Hasan, and temples has been noted by a number of scholars. Spencer, for example, says of Middle Kingdom rock-cut chapels such as those at Beni Hasan: "The placing of the shrine at the end heightened the axial nature of the plan and led to the development of a chapel resembling the form of a small

---

45 Steindorff, "Haus und Tempel," p. 107. See also Spencer, Lexicographical Study, pp. 20-23, for discussion of this term.

46 Scharff, "Das Grab als Wohnhaus," pp. 18-19.

47 Aldred, "Grabdekoration," LÄ II, col. 853; Kemp, Anatomy pp. 54-55.

48 See Reisner, Tomb Development.

49 Scharff, "Das Grab als Wohnhaus," p. 9.

50 See Badawy, Architecture I and II.

51 Junker, Giza II, pp. 105ff.; Scharff, "Das Grab als Wohnhaus," p. 46.

temple."[52] Badawy discusses analogies between the layout of rock-cut tombs such as the one studied here and later cult temples of the New Kingdom, which perhaps reflected Middle Kingdom examples which have not been preserved.[53]

The model of tomb as house has been prevalent in Egyptian scholarship on the subject of the decoration of tomb chapels. Scholars have long held that the primary purpose of decoration within elite tombs of the Old and Middle Kingdoms was to provide the deceased with the resources for a successful afterlife. According to this model, the tomb was the dwelling in which the deceased person would live forever, where they would enjoy an idealized version of the life which they had lived on earth.[54]

However, much of the decoration within elite tomb chapels does not fit neatly into this model, and requires further explanation. Using the alternative (but not exclusive) model of the tomb chapel as a sort of private mortuary temple may clarify both specific aspects of the chapel decoration and the function of the tomb chapel within the larger Egyptian cosmos.

Cultic Structures as Built Mirrors of the Cosmos

It has been demonstrated that a number of the types of monuments which form part of the built landscape functioned as models of the cosmos.[55] Within these microcosmoi, the ritual activities carried out would serve both to reflect and to guarantee the proper functioning of the cosmos. The cultic structure thus represented the entire process of cosmic maintenance: the structure existed within the Egyptian universe as a built element and at the same time acted to ensure its perpetual regeneration. This type of cosmic modeling has been seen in Old Kingdom pyramid complexes (which include both royal tomb and royal mortuary temple),[56] a royal ka chapel of the Middle Kingdom,[57] gods' temples of the New Kingdom and later periods,[58] and palaces of the New Kingdom.[59]

Of these structures, the god's temple of the New Kingdom and later periods and the royal ka chapel of the Middle Kingdom are the most useful with respect to the tomb of Khnumhotep II at Beni Hasan. The god's temple of

---

52 Spencer, Death in Ancient Egypt, p. 228.

53 Badawy, Architecture II, p. 195.

54 For example, E.B. Smith, Egyptian Architecture, pp. 87-89; W.S. Smith, HESPOK, p. xii; Aldred, "Grabdekoration," LÄ II, col. 856 and "Grabrelief," LÄ II, col. 868.

55 See references below.

56 Brinks, Die Entwicklung der königlichen Grabanlagen des Alten Reiches; Teichmann, Der Mensch und sein Tempel; Arnold, "Vom Pyramidenbezirk zum 'Haus für Millionen Jahre;'" Arnold, "Totentempel I and II," in LÄ VI.

57 O'Connor, "Nephepetre's Chapel at Dendereh."

58 Arnold, Wandrelief; Nelson, "The Egyptian Temple;" Baines, "Temple Symbolism;" Finnestad, Image; Hornung, Idea into Image.

59 O'Connor, "Mirror of the Cosmos: the palace of Merenptah."

the New Kingdom is useful because of its close architectural correspondence, and the royal ka chapel is crucial because of its relatively close temporal relationship.

Unfortunately, comparatively little remains of gods' temples of the Middle Kingdom, and little has been published on the interrelationship of architecture and decoration, which is in most cases difficult to study effectively from the remaining data.[60] The most extensively preserved god's temple of this period, at Medinet Maadi, is small and in very poor shape; what remains is a hypostyle hall whose ceiling is held up by two papyrus columns, a vestibule, and three sanctuaries in the rear.[61] Thus the basic ground plan resembles the axial layout of the New Kingdom temple, but only the decoration of the shrines was preserved, and therefore the overall decoration of this structure cannot be used for purposes of comparison.

There are several other religious structures of the Middle Kingdom, which would certainly be worthy of further study but which fall outside the scope of this work. Chief among these are the Satis Temple at Elephantine[62] and a series of well-known chapels at Karnak. Mention should also be made here of a mudbrick shrine dating to the time of Amenemhēt I at Ezbet Rushdi in the Delta, and Middle Kingdom remains, also in mudbrick, at the site of Medamud.[63] Again, at both sites, the remains are scanty and not particularly useful here as comparanda.

There are royal mortuary temples from both the Old and Middle Kingdoms, aspects of which provide an important link between the private mortuary temple and the god's temple as loci of worship for beings who have both earthly and divine natures. There are two main problems associated with using these structures for comparanda to the Beni Hasan tomb. First, even the reasonably standardized plan which is reached during the Old Kingdom and copied in the Middle Kingdom is considerably more complex than the plan of Khnumhotep II's tomb; extracting and analyzing the key elements of these temples is in itself a difficult task which would not be useful to outline here.[64] Second, very little of the decoration of these structures has been preserved *in situ*, thus it is difficult to use them for purposes of comparison.[65]

---

60 Cf. Kemp, Anatomy, pp. 64ff.

61 See Naumann, "Der Tempel des Mittleren Reiches in Medinet Madi."

62 See Kaiser et al., "Stadt und Tempel von Elephantine; 15./16. Grabungsbericht," pp. 152-57, with references to earlier reports.

63 See Kemp, " Old Kingdom...," p. 104 and Kemp, Anatomy, pp. 66-69.

64 See Arnold, "Vom Pyramidenbezirk," pp. 1-8 for a recent analysis.

65 The temples containing the best preserved decoration are of Pepi II (see Jéquier, Pepi II) and Sahure (see Borchardt, *Sa^{3} hu-re^{c}*).

The God's Temple of the New Kingdom and Later

The primary function of the god's temple was to ensure the proper continuance of the cosmic order by providing an effective environment in which the divine being was both empowered and placated, and in turn empowered the king as his or her deputy on earth.[66] The decoration of a god's temple focuses on the mutually beneficial relationship between the king and the god or gods worshipped there. The king, as chief celebrant, acts to empower and placate the god through various rituals and offerings and is given in return the wherewithal to rule in the earthly realm. The rule of the king is then represented in a number of ways, with emphasis on his role as guarantor of the cosmic order as expressed in the earthly realm, and thus as proxy on earth for the creator god.[67] The temple is the place in which the divinity was cared for through dogmatically defined communication, on which the welfare of the country and even the continuance of the world order depended. The decoration of the temple therefore served both to express and maintain the rites carried out in the temple.[68]

Figure I.6: Flow of Benefits in the God's Temple

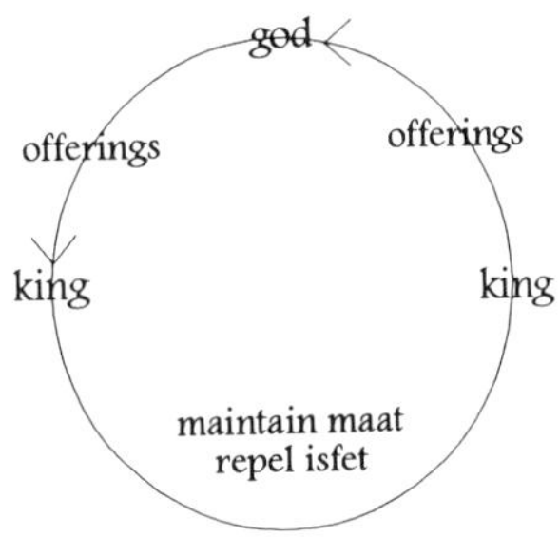

Careful analysis of the god's temple has led to the recognition of this type of structure as a model of the cosmos. This concept was first expressed by Rochemonteix in 1887,[69] and has since been expanded and clarified by a number of scholars.[70] The temple as microcosm is expressed through architectural and decorative features which identify it with the cosmic landscape; at least in the case of the Ptolemaic temple at Edfu, this cosmic imagery is also spelled out by many of the texts inscribed on the walls, some of

66 Cf. Gundlach, "Tempelrelief," LÄ VI, cols. 407-11.

67 Baines, "Legitimacy of Rule."

68 Arnold, Wandrelief, p. 3.

69 Rochemonteix, M. de, "Le temple égyptien," in Revue internationale de l'enseignement, 15 July 1887. Reference from Baines, "Temple Symbolism."

70 For example, in Nelson (1944), "The Egyptian Temple;" Reymond, The Mythical Origin; Baines (1976), "Temple Symbolism;" Finnestad (1985), Image; Mostafa (1989), "Reflexions sur la function cosmographique."

which identify various features of the temple with features of both local and universal cosmological geographies.[71]

The most clearly evident cosmographic features of a typical New Kingdom or later temple were as follows. The pylon was shaped in obvious reference to the hieroglyph for horizon, *akhet (ꜣḫt)*, above which the sun would rise (especially in the many cases where the temple was oriented East to West) each morning and set in the evening. The pylon and the outer walls bore images of the king vanquishing the enemies of Egypt, creating both an image of the outside world and an apotropaic boundary between the profane world outside the temple precinct and the sacred realm within.[72] The outer, or festival court, can perhaps be seen as the orderly cosmos, the open space in the center being equivalent to a body of water, perhaps even the river, surrounded by a portico supported by vegetal columns, representing the surrounding land.[73]

The interior of the temple can be said specifically to represent the landscape of cosmogony, in its watery and nocturnal state.[74] The floor of the hypostyle hall represents the inundated earth, out of which grow columns in the form of marsh plants, usually lotus and papyrus. Papyrus and lilies could be carved on the dado around the hall, and also on the bases of the columns, which themselves border the forecourt as if they were plants around a pond or rising from inundated earth; these columns appear to grow toward the sky. A blue line could even be painted on the base of the columns, indicating the water level.[75] The nocturnal aspect of the cosmos is expressed through the ceiling, which is painted blue and covered with golden stars, which identify it with the night sky. This identification is emphasized through the darkness of the temple interior, into which little light is permitted. This darkness intensifies as one moves toward the inner sanctuary.[76]

The temple thus represents the emergent cosmos. This is also expressed by a gradual raising of the floor of the temple from the outer court into the inner sanctuary. The sanctuary corresponds to the primeval mound, the first earth to emerge from the waters of Nun, on which the creator god stood to initiate the Creation.[77] The image of the god stood in this shrine, and was thus identified with the creator god. The inner sanctuary represents the center of the cosmos; in ritual texts, it is stated that the god sleeps here with the doors closed, underlining the concept of night.[78]

---

71 Finnestad, Image, pp. 3-5.

72 Baines, "Temple Symbolism," p. 10.

73 For the festival court, David O'Connor, personal communication.

74 Baines, "Temple Symbolism," p. 12.

75 Borchardt, Pflanzensäule, pp. 53ff.

76 Finnestad, Image, pp. 11-13 and note 37.

77 Barguet, Le temple d'Amon-Rê, p. 332, note 1; Baines, "Temple Symbolism," pp. 10-11.

78 Finnestad, Image, pp. 11-13 and note 37.

Each day, the doors to the temple were opened and light was let in. Down the long axis of the temple ceiling are often found representations of winged sun disks, so that the long axis represents the path of the sun which travels into the shrine, whose doors were opened last.[79] When the shrine doors were opened, the light reflected from the divine image, mirroring the moment of Creation. Thus the temple is the place where the transition from Chaos to created Cosmos is re-enacted continuously.[80] This transition is not necessarily a peaceful one, and the creator god is seen as a warrior god who establishes the cosmos through effort. This continuous conflict is reflected in the Ptolemaic temple of Horus at Edfu by motifs such as the subduing of the enemy and the battles of Horus against various enemies,[81] and in general by images of the king vanquishing his foes.

There are a number of cosmogonic texts associated with the temple of Horus at Edfu. Horus, as would be expected here, is the creator god, and cosmogonies of both the cosmos in general and Edfu in particular are provided; ultimately, the building of the temple as the final goal of Creation is described.[82] The temple is, therefore, the result of Creation, and at the same time constitutes ensurance that the perpetual regeneration of the cosmos will be effected.

Finnestad also has shown that the temple of Edfu at least, and possibly other temples of this period and earlier, can be identified not only with the emergent Cosmos, but with the land of Egypt as a whole and with the immediate geography of the relevant area.[83] For example, the Edfu temple is identified with the land of Egypt through the term *tawy (tꜣwy),* meaning the Two Lands, the regions of Upper and Lower Egypt. Another name for the Edfu temple is *akhet* (horizon), which connotes the farthest limits of the country. This name is also applied specifically to the temenos wall, as the outer limit of the temple, or to the pylon of the temple, where the transition from the outside world into Egypt and the confrontation between the two is emphasized.[84]

The Royal Ka Chapel of the Middle Kingdom

A similar interpretation of a small chapel of the Middle Kingdom at Dendereh, dedicated to Nebhepetre Mentuhotep, has recently been proposed by O'Connor. Through detailed analysis of the decoration of this chapel within its architectural framework, he has shown that the scenes within this chapel function at a number of levels: they form a narrative which relates both the

---

79 Nelson, "Egyptian Temple," pp. 47-48; Arnold, Wandrelief, p. 3.

80 Finnestad, Image, pp. 12-14.

81 Finnestad, Image, pp. 14-16.

82 Finnestad, Image, pp. 24ff.

83 Finnestad, Image, p. 10.

84 Finnestad, Image, pp. 8-10.

story of Nebhepetre himself and of Egyptian kingship in general; at the same time, they reflect the cosmogony, cosmography, and cosmic process.[85]

OVERVIEW

This study will investigate the issues discussed above as they relate primarily to the 12th Dynasty tomb chapel of Khnumhotep II at Beni Hasan. In Chapter II, the general background for this work is presented. The political situation in early Dynasty 12 is briefly summarized. The geographical situation of Beni Hasan within Egypt, and the locations of important towns and cemeteries within the nome as reconstructed from texts and archaeological remains, are discussed. Finally, the basic features of Khnumhotep II's tomb are described.

The methodology with which the analysis of Tomb 3 was prepared is presented in Chapter III. Included with the outline of this approach is the previous scholarship on elite tombs which has provided the basis of my research.

Chapter IV presents a detailed analysis of the decoration of Khnumhotep II's tomb chapel. The decoration of each wall is broken down into microscenes, defined as groups of figures which are closely related through the activities which they perform and/or the physical setting in which they are depicted. Each of these individual microscenes is briefly described, and the various relevant levels of meaning contained within it are discussed. The microscenes are then considered within the context of each wall as a whole. (The shrine, which has decoration on four walls, is treated as a single entity here for reasons of convenience.)

In Chapter V, the analysis is taken further: the information presented in Chapter IV is synthesized, and the organization and function of the decoration of the chapel as a whole are elucidated through analyses of the decoration within the chapel as a whole. Coherent reasons for the deliberate placement of various decorative elements in particular locations within the monument are suggested, and overarching themes portrayed by the decoration are mapped within the tomb. The detailed analysis of the decoration of the tomb chapel of Khnumhotep II at Beni Hasan presented in this volume reveals multiple layers of meaning and function similar to those found in other forms of Egyptian religious architecture, and shows that this chapel, like a temple, can be viewed as a model of the Egyptian cosmos on several levels.

---

85 O'Connor, "Nebhepetre's Chapel."

In Chapter V, Tomb 2 at Beni Hasan is briefly described, analyzed, and compared to Tomb 3; this tomb is also found to function as a miniature cosmos and a private mortuary temple.

CHAPTER II :
BACKGROUND:
HISTORY, GEOGRAPHY, AND ARCHITECTURE AT BENI HASAN

In this chapter, the geographical and historical settings of the 12th Dynasty tomb of Khnumhotep II at Beni Hasan are discussed, and the architecture and general layout of the tomb are described briefly.

GENERAL HISTORICAL BACKGROUND OF EARLY DYNASTY 12

The period of the early 12th Dynasty was a time of rebuilding for Egypt. Amenemhēt I, who was a vizier under the last king of the 11th Dynasty, Mentuhotep IV, usurped the throne and moved the capital of the country from Thebes, where the 11th Dynasty had been based, to the Faiyum entrance at Itjy-tawy. The 12th Dynasty rulers built large pyramids for their tombs at Lisht and Dahshur, following the architectural example of the Old Kingdom. In this, as in many things, the kings of the 12th Dynasty considered themselves the successors of the powerful and highly-centralized rulers of the Old Kingdom. Military expeditions were made into Nubia, where a string of fortresses was built to maintain control over what were now essentially occupied territories; fortresses were also built along the northwest borders of the Delta, and there is evidence for campaigns into Asia.[1] The country was once again prosperous

[1] See Kemp, "Old Kingdom...," pp. 141-143.

and the government centralized and powerful, as attested by the numerous remains of extensive building projects.[2]

Figure II.1: Map of Egypt

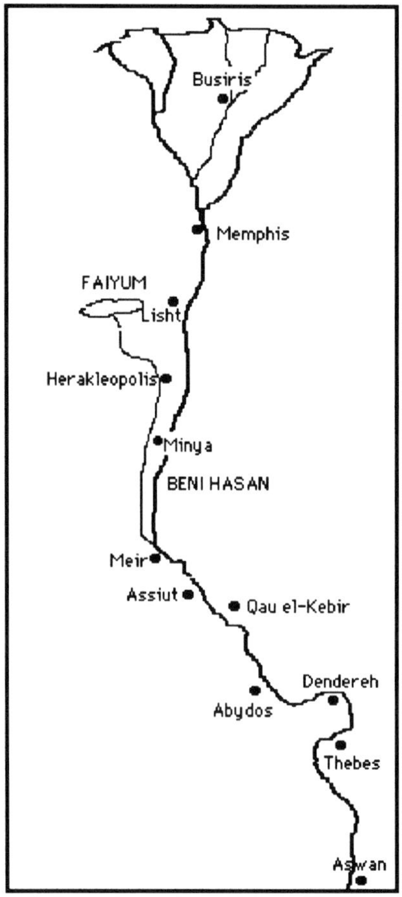

There is evidence that the governmental bureaucracy was re-organized by Amenemhēt I. Beginning in his reign, provincial governors were set in place, or at least confirmed, by the king. This is certainly the case at Beni Hasan, where Khnumhotep I of Tomb 14, Ameni of Tomb 2, and Khnumhotep II of Tomb 3 make frequent reference to their respective sovereigns. This is in contrast to the earlier tombs at this site, in which kings are never named.

---

[2] For general treatments of this period, see Gardiner, Egypt of the Pharaohs; Kemp, "Old Kingdom, Middle Kingdom, and Second Intermediate Period," in Social History; and Hayes, "The Middle Kingdom in Egypt," in The Cambridge Ancient History.

It has traditionally been thought that the administration of Egypt was reorganized under Senwosret III, who was thought to have taken power from the provinces and abolished the title of Nomarch,[3] returning the responsibility for the provincial nomes to the central government.[4] It is no longer widely believed that the changes in the provinces seen from the reign of Senwosret III on are necessarily due to a curbing of provincial power, since there appears to be at least one set of very large tombs from this period, at Qau el-Kebir.[5] In any case, the last attested Nomarch at Beni Hasan is Ameni, who lived at least into the early reign of Amenemhēt II; the last large inscribed tomb was built for an Overseer of the Eastern Desert, Khnumhotep II, who may have lived into the reign of Senwosret II.

GEOGRAPHY OF THE BENI HASAN AREA

Beni Hasan is located in Middle Egypt, within the ancient boundaries of the 16th Upper Egyptian nome, called the Oyrx nome. This region was bordered on the north by the 17th U.E., or Hare Nome and to the south by the 15th U.E., or Jackal nome.[6] In the Old and Middle Kingdoms, the Oryx nome encompassed about 25 kilometers (16 miles) along the length of the Nile, stretching from the modern towns of Sheikh Mubarak in the north to Itlidem in the south (see Figure 2.2). The site of Beni Hasan lies about 23 kilometers (15 miles) south of the modern town of Minya, on the east bank of the Nile.

The Nile runs slightly southeast-northwest through much of the Oryx nome, although it is oriented almost due north near the southern boundary. In the south of the nome, where Beni Hasan is located, the river also widens considerably, leaving a small island between its two arms. The site of Beni Hasan lies just before this widening. The cliffs of the eastern high desert approach the river closely here, leaving only a thin strip of fertile land and a narrow band of low desert. The eastern cliffs may have been chosen for the tombs because the river runs so close to them; on the west bank, the desert is far from the Nile, thus tombs there would have been considerably more difficult to reach from the river. The tombs command an impressive view across the river, to the extensive cultivated land which lies on the opposite bank.[7] Although the course of the river may have differed somewhat in

---

[3] The last attested "Great Chief" is Djehutyhotep II of Bersheh, dating from the time of Senwosret III (see Brovarski, "Ahanakht," p. 29).

[4] Hayes, "The Middle Kingdom," pp. 505ff.

[5] See Quirke, Administration, p. 5, note 3.

[6] Baines and Málek, Atlas, pp. 14-15.

[7] Kessler, Topographie, p. 130. Similar views are to be had from the tombs at Meir and Deir el-Bersheh.

antiquity,[8] the modern terrain still reflects the landscape of antiquity fairly accurately.

Figure II.2: Sketch Map of the Oryx Nome

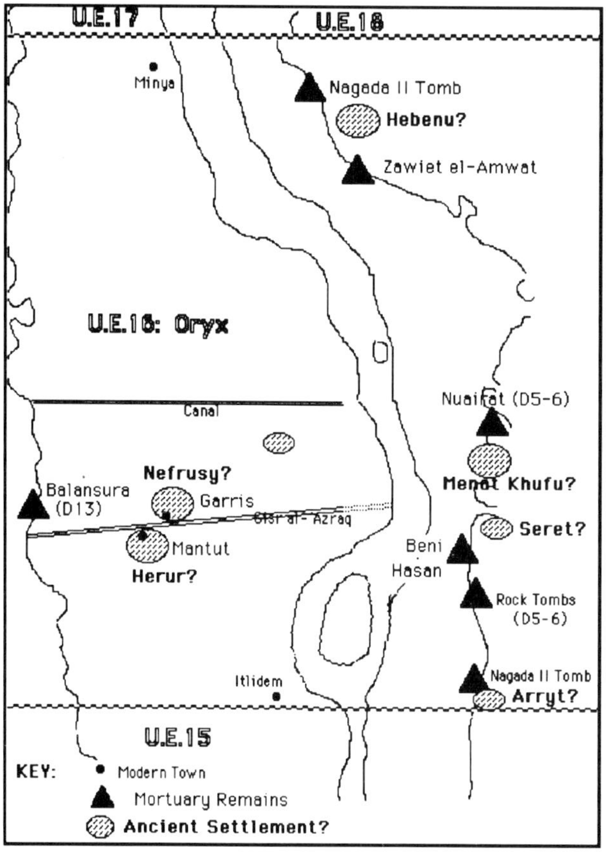

MORTUARY REMAINS

A number of mortuary sites have been located within the boundaries of this nome. In the northern part of the nome, on the eastern bank, is the site of Zawiet el-Amwat,[9] at which tombs of the late Predynastic Period, an unfinished pyramid of the 3rd Dynasty, and several late 6th Dynasty elite tombs have been found.[10]

---

[8] It is possible that the Nile was farther to the west during ancient times (Butzer, Hydraulic Civilization, pp. 34-35).

[9] Also called Zawiet el-Meitin.

[10] Porter and Moss, Topographical Bibliography IV, pp. 134ff.

The eastern cliffs in the southern part of the nome shelter several cemeteries of rock-cut tombs, dating from the Old Kingdom through the Late Period.[11] The site of Beni Hasan lies just east of the modern village of the same name. There are two cemeteries here: the upper range, which includes the tomb of Khnumhotep II, and the lower necropolis. The lower cemetery first came into use in the late Old Kingdom, with the construction of a group of simple chamber tombs and "pit tombs" in the lower slopes which lie at the base of the cliffs. The walls of these tombs are sometimes inscribed; the names of the owners and the pottery suggest dates primarily in the 5th and 6th Dynasties.[12] These tombs were frequently re-used in the Middle Kingdom, as evidenced by finds of later coffins and wooden models.[13]

In addition to Old Kingdom chamber tombs, the lower cemetery at Beni Hasan contains almost 900 shaft tombs dated by the excavator to the Middle Kingdom, each built with a small chamber at the bottom which held the coffin and the funeral deposits. The plethora of shafts, coffins and burial goods, which included numerous wooden models of granaries, boats, offering bearers, and the like,[14] suggested to Garstang a prosperous local mortuary industry which lasted for several hundred years.[15] The owners of the lower tombs would perhaps have been inhabitants of the nome and minor officials under the nobles buried in the upper rank of tombs. Garstang found chapels associated with only two of the shaft graves, one of which dates to the reign of Senwosret III, and thus follows the latest of the dated upper tombs.[16]

The tomb of Khnumhotep II is in the upper cemetery, one of thirty-nine high-status tombs cut into the cliff face high above the river. These tombs lie in a north-south row, divided into two groups by a slight break in the natural terrace onto which they open. Thirteen of the thirty-nine tombs were more or less extensively decorated with scenes and texts; inscriptions within these tombs tell us that they were built for members of the elite of the nome, ranging in rank from Nomarch of the Oryx nome[17] or Overseer of the Eastern Desert[18] to scribe. Khnumhotep II's tomb is labeled Tomb 3.

---

[11] Kessler, Topographie, pp. 190ff; Garstang, Burial Customs, pp. 26-27.

[12] Garstang, Burial Customs, pp. 15 and 30-34.

[13] See especially Tomb #481 of Dynasty 6 (Garstang, Burial Customs, pp. 36-41) and Tomb #482 (Garstang, Burial Customs, pp. 42-43).

[14] See for example, Garstang, Burial Customs, pp. 56-57 and pp. 65ff.

[15] Garstang, Burial Customs, pp. 45-48.

[16] Garstang, Burial Customs, pp. 51-53.

[17] *ḥry-tp ꜥꜣ n Mꜣ-ḥḏ*, cf. Fischer, "Gaufürst," LÄ II, col. 413.

[18] *imy-r ḫꜣswt iꜣbtt*.

The following chart lists the principal decorated tombs in the upper range at Beni Hasan:[19]

Table II.1: Decorated Tombs at Beni Hasan

#	NAME	MAJOR TITLES	FATHER	MOTHER	WIFE	CHILDREN	OTHER INFORMATION
2	Amenemhēt (Ameni)	Nomarch of Oryx; Hereditary Prince.; Overseer of priests of Khnum.	?; was Hereditary Prince; Generalissimo of Oryx	Henu (daughter of a Count)	Hetept /Hathorhetep (daughter of a Count)	Khnumhotep	Bio dated to year 43 of Senwosret I, corresponding to year 25 in the Oryx Nome.
3	Khnumhotep (II)[20]	Overseer of the Eastern Desert; Hereditary Prince; Count of Menat Khufu; Overseer of priests	Neheri (son of Sebekankh, Hereditary Prince, Ruler of the New Towns)	Bakt (daughter of Khnumhotep I and Satip?)	Khety (daughter of a ruler of the Jackal nome and Tjent); Tjat	By Bakt: Nakht; Khnumhotep; Neheri; Neternakht; Bakt; Tjent; Meres.By Tjat: Neheri; Khnumhotep (IV); Satip	From bio: Tells how his mother's father (prob. K I of Tomb 14) was appointed Overseer of ED in Menat Khufu and then Nomarch of Oryx by Am. I. K I (?) appoints his son, Nakht (K II's uncle), as a ruler in Menat Khufu. Am. II appoints K II Hereditary Prince, Overseer of the Eastern desert, and to the inheritance of mother's father (unnamed, but probably K I) in Menat Khufu in Year 19.
4	Khnumhotep (III)		(Khnum-hotep II)	Tjat			only portico was finished
13	Khnumhotep	royal scribe; various religious titles	Netjerhotep	Sattekh			

[19] Newberry numbered the Beni Hasan tombs consecutively from North to South, and each tomb is traditionally referred to by its number.

[20] Porter and Moss refer to the owner of Tomb 3 as Khnumhotep III, counting a man of lesser status (a scribe) who was buried in Tomb 13 as Khnumhotep II (Topographical Bibliography IV, p. 149.) (Tomb 13 was built for this man.) Following the more common usage, I have given numbers only to men of highest rank, thus Tomb 14 belongs to Khnumhotep I and Tomb 3 to Khnumhotep II.

#	NAME	MAJOR TITLES	FA-THER	MO-THER	WIFE	CHIL-DREN	OTHER INFOR-MATION
14	Khnum-hotep (I)	Nomarch of Oryx, Count of Menat Khufu; Overseer of priests	?	Bakt	Satip	Nakht; Bakt (named in Tomb 3)	bio: hard to read, mentions Am. I and a naval expedition
15	Bakt III	Nomarch of entire Oryx	Remu-shenti	Hetep-erfu		Hathor-nefer-heputa	
17	Khety	Nomarch of entire Oryx; Overseer of the Eastern Desert?[21] Captain of Soldiers in all difficult places	Bakt (III?)		Khnum-hotep	Khety	
21	Nakhti	Nomarch of Oryx; Overseer of Eastern Desert					only decoration left is part of scene on south wall.
22	un-finished						false doors with cavetto cornices in the middle of the north and south walls; were ori-ginally painted
23	Netjer-nakht	Overseer of Eastern Desert; Hereditary Prince; Overseer of priests of Horus, Smiter of the Rekhyt.		Ary-hotep	Herib		Khnumhotep son of Neheri (K II) claims to have built this for his ancestor (*it*); probably was his immediate suc-cessor, based on royal idea that next in line buries pre-decessor.
27	Remu-shenti	Nomarch of Oryx					
29	Bakt I	Nomarch of Oryx			Tutqa	Bakt (II)	
33	Bakt II	Nomarch of Oryx	Bakt I	Tutqa			

There is currently some debate as to both the sequence and the absolute dates of the large, inscribed rock-cut tombs at Beni Hasan. The earliest tombs have been dated at one extreme to the end of Dynasty 11 (c. 1991),[22] and at the other to Dynasty 9 (c. 2160).[23] Supporters of the late chronology account for the presence of a relatively large number of tombs belonging to "Nomarchs" in a

---

21 Newberry suggests this translation of some of the fanciful glyphs on the north wall of this tomb, see BH II, Pl. 14 and p. 58.

22 Schenkel, Frühmittelägyptische Studien, pp. 79-84; Hölzl, Felsengräber, pp. 31ff. and presentation at Metropolitan Museum, 1991.

23 Spanel, Beni Hasan in the Herakleopolitan Period; see also Spanel, Two Studies, forthcoming.

relatively short period of time by hypothesizing the existence of multiple contemporaneous holders of this title.[24] However, analysis of epigraphic, architectural, iconographic, and genealogical elements in these tombs suggests that while such a situation is possible, it is not necessary or even likely.[25] However, there is no serious question about the dating of Tomb 3, since it contains specific regnal dates of the 12th Dynasty kings Amenemhēt II and Senwosret II. According to this evidence, therefore, this tomb was built and decorated between c. 1910 and c. 1891 B.C.[26]

In the 13th Dynasty, the principal cemetery of the nome moved to the west bank, to the area of modern Balansura, which lies in the low desert bordering the cultivated fields. This cemetery remained in use into the 18th Dynasty. On the east bank of the river, the cemeteries lay unused from the 12th Dynasty until the 20th Dynasty and later, when more tombs were carved in the area of the Wadi Batn al-Baqara.[27]

SETTLEMENTS

There is very little archaeological evidence for Old and Middle Kingdom settlement sites within the boundaries of the Oryx nome.[28] The New Kingdom rock-cut temple of Speos Artemidos, in the Wadi Batn al-Baqara, is thought to have replaced a Middle Kingdom temple of Pakht.[29] Apart from this structure, only isolated finds near modern sites and town mounds suggest that certain modern villages lie on the sites of ancient mounds. Textual attestations, both from the tombs in the area and from non-local sources, provide evidence for the existence and locations of a number of important towns which may have been located within the Oryx nome.[30]

Table II.2 summarizes the towns mentioned in the upper tombs at Beni Hasan, generally as the homes of various divinities. As this chart demonstrates, the town named earliest and most consistently by the Beni Hasan nobles is Herur, sacred to Khnum, which appears in almost every elite tomb that retains a significant amount of decoration. It does not appear in

---

24 Schenkel, Frühmittelägyptische Studien, pp. 79-84; Hölzl, Felsengräber, p. 45, Table 2.

25 For discussion of the earlier chronology, see Spanel, Beni Hasan in the Herakleopolitan Period, which focuses mainly on the epigraphy and paleography of the inscriptions; see also Spanel, Two Studies, forthcoming. A detailed study of the decoration of all of the tombs has provided chronological information which suggests that Newberry's reconstruction of the sequence of the tombs, to be found in BH I, p. 3, is still useful, and is probably correct for the most part (although Tomb 13 is probably not the earliest of the tombs).

26 In his biography (lines14-20), Khnumhotep II records his attainment of the titles of Overseer of the Eastern Desert and Count of Menat Khufu in year 19 of Amenemhēt II (c. 1910 B.C.); an inscription on the north wall records a date of year 6 of Senwosret II (c. 1891 B.C.).

27 Garstang, Burial Customs, pp. 17-18; Kessler, Topographie, p. 63.

28 Kessler, Topographie, pp. 130-34.

29 Kessler, Topographie, p. 63, see footnote 279.

30 See Gardiner, Ancient Egyptian Onomastica II, 28, 82ff., 85ff.; Kessler, Topographie.

Tomb 3, the tomb of Khnumhotep II, perhaps because he was not a Nomarch of the Oryx Nome, but was instead an Overseer of the Eastern Desert. The town of Nefrusy, whose mistress was Hathor, begins to appear in the middle group of tombs, and is mentioned by almost every tomb owner from this time forward.

The locations of these various towns are assumed to be within the borders of the nome. Primarily on the basis of textual evidence, a number of scholars have suggested sites for these towns; most recently, Kessler has produced an exhaustive survey of the area of the 15th-17th Upper Egyptian nomes in which, after reviewing all the literature on the subject and examining all of the available textual and topographic clues, he provides reasonable guesses as to the most likely locations for these towns.[31]

Table II.2: Towns mentioned in the Beni Hasan Tombs[32]

TOMB #	HERUR	NEFRU-SY	MENAT KHUFU	HEBE-NU	IW-BU	ARRYT	SERET
29	x						
33							
27	x						
15	x	x					
17	x	x					
14	x	x	x	x	x		
2	x	x		x	x		
3		x	x				x
23						x	

Kessler believes that Herur and Nefrusy were both located near one another in the western floodplain, perhaps along the modern-day Gisr al-Azraq (which may itself reflect the course of an ancient canal).[33] Nefrusy was an extremely important city during the Second Intermediate Period;[34] it is likely that it was the capital of the nome in the Middle Kingdom.[35] Herur was

---

31 Kessler, Topographie; see also his Lexikon der Ägyptologie articles on these towns. His approximate locations for the various sites are indicated in Figure 2.1.

32 The tombs in this table are arranged chronologically.

33 Kessler, Topographie, pp. 157-59 and178-85. Alternative locations are suggested by Baines and Málek (Atlas, p. 84 and # 379); however, according to their map, the sites of both of these towns fall in the territory of the 15th Upper Egyptian nome.

34 See Smith and Smith, "Kamose Texts," p. 60.

35 Kessler, Topographie, pp. 136-37.

presumably the site of an important temple of Khnum,[36] perhaps in conjunction with Hekat, while Nefrusy contained a temple of Hathor. Kessler's reconstruction places these sites almost due west of Beni Hasan.

The town of Menat Khufu is cited only in securely dated 12th Dynasty tombs, and then only in tombs belonging to men with the title "Overseer of the Eastern Desert" (Tombs 14 and 3). This town is thought to have been on the east bank, north of Beni Hasan, in a position controlling a major wadi into the eastern high desert.[37] Newberry suggests that this city was the capital of the eastern part of the nome, under the control of the Overseer of the Eastern Desert[38] (who also held the title of Count of Menat Khufu). Kessler also suggests that a temple to Horus, Smiter of the Rekhyt, a god mentioned at least once in each of the five best-preserved tombs in the upper range, was located here.[39] Khnumhotep II of Tomb 3 was a priest of this god, and Bakt I of Tomb 29 built a monument for him, so there must have been a temple somewhere in the area. Menat Khufu is as likely a site as any. This town is not mentioned in any post-Middle Kingdom sources, thus Kessler hypothesizes that its name might have been changed.[40]

Seret, sacred to the goddess Pakht, is the only other town mentioned in Tomb 3. Since the temple of Pakht of Seret has been identified with the Speos Artemidos, Kessler tentatively places it on the east bank, just north of Beni Hasan.[41] However, an associated settlement has not been found.[42]

GENERAL DESCRIPTION OF TOMB 3

The information on the layout and decoration of the tomb of Khnumhotep II which forms the basis for this study was gathered primarily from the principal publication of Beni Hasan. This work consists of four volumes produced by Percy E. Newberry (under the auspices of the Egypt Exploration Fund) from 1893 to 1900.[43] Newberry's publication provides plans

---

[36] Cf. Newberry, BH II, p. 20. Khety of Tomb 17 records making "a monument of eternity in the temple of Khnum, lord of Herur;" and Ameni of Tomb 2 was "Overseer of the God's Servants of Khnum of Herur."

[37] Kessler, "Menat-Chufu," LÄ IV, col. 41; Newberry, BH II, pp. 18-19. Others have suggested that Menat Khufu was the capital of the entire nome (Helck, "Gaue," LÄ II, col. 391), and located it on the west bank, for example at Minya (see, for example, Baines and Málek, Atlas, p. 14).

[38] Newberry, BH II, pp. 18-19.

[39] Kessler, Topographie, p. 135.

[40] Kessler, Topographie, p. 135.

[41] Kessler, Topographie, p. 127.

[42] Kessler, Topographie, p. 134.

[43] Beni Hasan, volumes I to IV. References to copies made by early travelers to the site can be found in Newberry, BH I, pp. 10 and 40; Porter and Moss, Topographical Bibliography IV, pp.

and sections of architectural layouts, black and white copies (generally lacking in interior detail) of most scenes and texts, and brief written summaries of each of the tombs. In addition, I had access to a useful series of slides, taken by Melissa Robinson, Stacie Olson, and myself during a visit to Beni Hasan in 1988.[44] Photographs of Tomb 3 were also provided to me by Judy Lustig, who visited the site in 1991; I am most grateful to her for the use of this material.

Tomb 3, the tomb of Khnumhotep II, is the latest in the series of large inscribed tombs at Beni Hasan. It lies toward the north end of the series of tombs, and is carved straight back into the cliff so that its "east-west" axis actually runs slightly north of due east.[45] Royally dated inscriptions and references within the tomb indicate that the tomb owner held his principal position in the nome from Year 19 of Amenemhēt II and probably into the co-regency of this king with Senwosret II (c. 1910 - 1891 B.C.). The tomb was presumably built and decorated during this time.

This tomb is of the so-called "portico-in-antis" type, with an east-west central axis.[46] A long "dromos," now recognizable as a simple path lined with boulders, runs from the edge of the cultivation up a scree slope to an open courtyard bounded on three sides by low stone walls. The presence of this path suggests that there might once have been a valley chapel associated with this tomb.[47] Such chapels are preserved as part of several late Middle Kingdom tomb complexes at Qau el-Kebir,[48] and reference to what is presumed to be a valley chapel is made in the contracts of Djefaihapi at Assiut.[49]

On the west side of the courtyard, directly in front of the tomb entrance, is a small roofed portico, supported by two multi-faceted columns. An architrave spans the space above the two columns, and continues to the side walls of the portico, where it ends in two plain rectangular engaged pillars which project from the walls. The roof of the portico from the architrave to the tomb façade is vaulted, with its long axis running north to south; the west edge of this roof has been carved with representations of wooden beam ends.

A rectangular doorway with a raised threshold cuts through the cliff façade at the back of the portico. The rock faces around this opening have been

---

141-149. For more recent publications of features of these tombs, see Spanel, Beni Hasan in the Herakleopolitan Period, and Hölzl, Felsengräber; for color photographs of some of the paintings, see Shedid, Felsgräber.

44 The camera, film, and most of the expertise for this project was provided by Melissa Robinson.

45 However, all orientations, unless otherwise stated, will be approximated as if the tomb's axis were directly east-west.

46 The following description is adapted primarily from Newberry, BH I, pp. 52-53 and Pl. 22; and Newberry, BH II, pp. 72-75.

47 Badawy, Architecture II, p. 128.

48 Badawy, Architecture II, pp. 151ff.

49 Reisner, "Hepzefa," p. 90.

smoothed, painted to represent rose granite, and carved with inscriptions. Above the doorway is a list of festival days on which invocation offerings were to be made to Khnumhotep II; on the jambs are prayers to Osiris (north) and Anubis (south), each above a seated figure of Khnumhotep II facing inward. Pleas to visitors for invocation offerings were carved on the thicknesses of the entrance.[50] A pivot hole on the threshold indicates that this doorway once contained a door that opened inward.

Figure II.3: Plan and Elevation of Chapel of Tomb 3

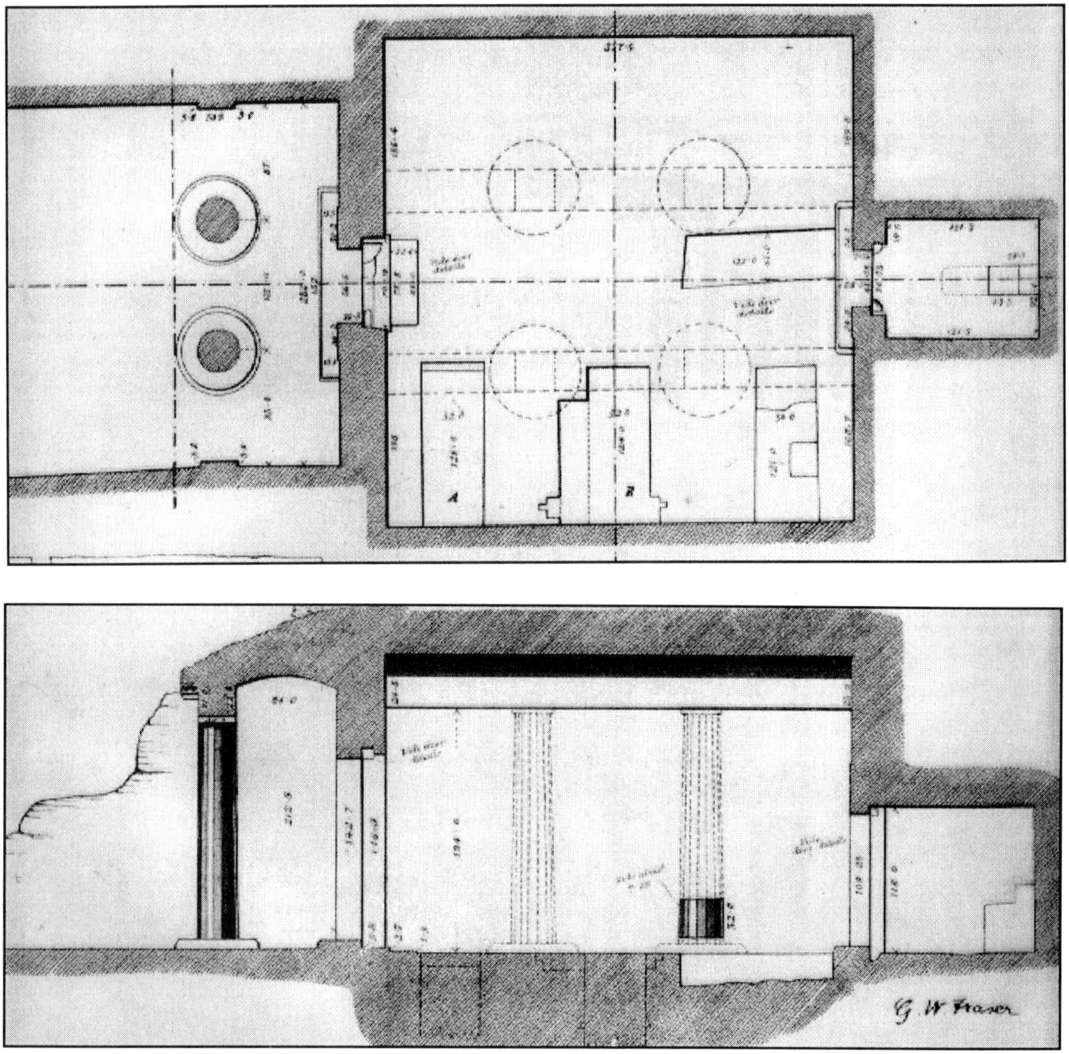

[50] For the architectural texts in this tomb, defined as texts associated primarily with architectural features, see Newberry, BH I, Pl. 24.

MAIN CHAMBER

From the entrance doorway in the west wall, three steps lead down to the floor level of the main chamber of the tomb chapel, which measures approximately 9.6 meters east to west by 9.7 meters north to south. This chamber is divided into three sections by two rows of columns surmounted by east-west architraves. Only one fragment from any of these columns was recovered; it shows that they were 16-sided and tapered slightly toward the top. The architraves are each inscribed with a request for invocation offerings, and a barrel vault roofs each of the three sections into which the architraves divide the ceiling. Vaults are first seen in Early Dynastic tombs at Saqqara, and are common in both brick and stone in the funerary architecture of the Old and Middle Kingdoms.[51] It has also been suggested that brick vaults were also used in houses, such as those at Illahun.[52]

The ceiling vaults in Khnumhotep II's chapel were decorated with alternating red and yellow squares containing black and blue quatrefoils, producing a pattern that suggests textiles.[53] These patterned vaults may make reference to the architecture of the "tent shrine," derived from a temporary, portable shelter built of wood covered with woven matting. Kemp hypothesizes that such structures were originally designed to house portable images of deities, and were also used in royal contexts.[54] According to this model, the tomb would represent three of these shrines, placed side by side, with a fourth (the outer portico) placed across the front.[55]

Figure II.4: Reconstruction of Tent Shrine[56]

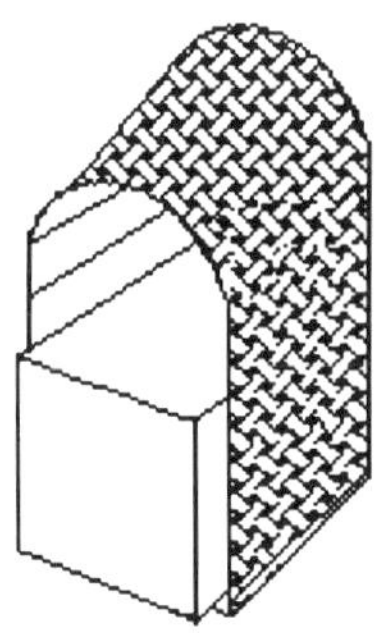

---

[51] Spencer, Brick Architecture, pp. 123-125; Arnold, Building in Egypt, pp. 186-201.

[52] Spencer, Brick Architecture, p. 125.

[53] Cf. Roik, Wohnhaus.

[54] Kemp, Anatomy, pp. 91ff.

[55] Note, however, that the beam ends carved along the western façade of the portico create a somewhat ambiguous architectural setting.

[56] After Kemp, Anatomy, figure 33.2.

The entrance to the tomb, the entrance to the shrine, and the dado of the main chamber have all been painted to imitate rose granite. This may be an attempt to create the illusion of more expensive, and possibly symbolically significant materials within the tomb. The lining of royal mortuary temples with granite is attested from the Old and Middle Kingdoms, and both granite and fine limestone are found casing the interior of burial chambers as early as the 2nd Dynasty.[57]

The walls of the main chamber were decorated with colorfully painted scenes. A khekher frieze above a band of colored rectangles runs around the tops of the paintings in both rooms; vertical borders of colored rectangles edged by "rope" patterns and finally narrow vertical blue lines frame each wall. The decorative elements that can be found framing the scenes and ornamenting various architectural features such as architraves and ceilings are thought to emphasize the identification of the tomb with a house.[58] They make reference generally to building in less permanent materials, following the example set in the 3rd Dynasty funerary monument of Djoser at Saqqara, in which many perishable materials have been translated into stone.[59] For example, khekher friezes such as are found in this chapel are thought to be modeled after the tops of reed mats, where the ends of the reeds have been bundled together.[60]

Above a low base of imitation rose granite, the dado in the shrine has been painted to look like niches covered with textiles. A similar pattern is seen on one wall in the 6th Dynasty burial chamber of the pyramid of Unas at Saqqara, which might suggest that earlier burial chambers were actually covered with woven mats.[61] This feature in turn is reminiscent of "palace façade" niching such as is seen on the enclosure walls of Early Dynastic "fortresses" at Abydos, and on the outer faces of Early Dynastic tombs at Saqqara.[62] This pattern is also seen on sarcophagi from the Early Dynastic Period through the Middle Kingdom,[63] and is part of the serekh in which royal names can be written.

AUTOBIOGRAPHY

Two horizontal bands of color, red above and yellow below, run directly beneath the scenes in the main chamber. Below these bands, on the "granite" dado, is a long autobiographical inscription, which was first carved

---

57 Arnold, Building in Egypt, p. 170.

58 Aldred, "Grabdekoration," LÄ II, col. 856.

59 Edwards, Pyramids, pp. 51ff.

60 Aldred, Egyptian Art, p. 52.

61 Piankoff, Unas, Pl. 7.

62 Emery, Archaic Egypt, pp. 175ff.

63 Spencer, Death in Ancient Egypt, pp. 166-174.

and then painted green.[64] The autobiography begins to the left of the entrance to the shrine on the east wall and runs counterclockwise around the walls of the main chamber, ending to the right of the shrine doorway.[65] There is a small false door in the middle of this inscription, along the south wall.

Figure II.5: Location of Autobiography[66]

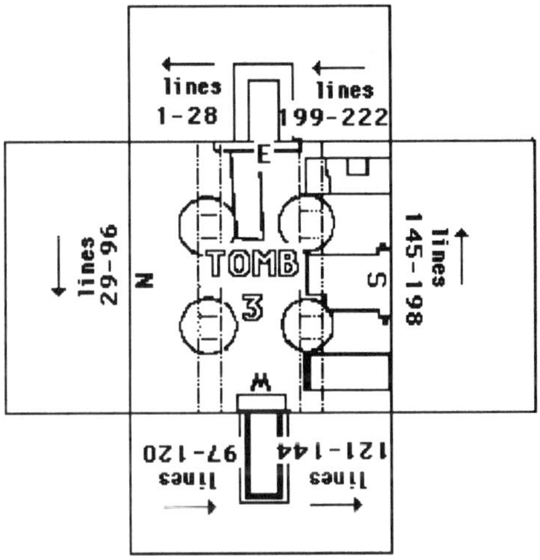

Two types of information are contained in this autobiography: historical data, and descriptions of the activities of Khnumhotep II. The former describes the life of the tomb owner's maternal grandfather, whose name is not given but who is clearly to be identified with Khnumhotep I of Tomb 14.[67] This man became Overseer of the Eastern Desert[68] first and then Nomarch of the Oryx Nome[69] later, under Amenemhēt I; he was followed in these positions by his eldest son, Nakht (Tomb 21).[70] Khnumhotep II himself succeeded to the title of Overseer of the Eastern Desert in year 19 of Amenemhēt II[71] through his

---

64 For translation, see Breasted, Ancient Records, II, 279-89; Newberry, BH I, pp. 57-66.

65 Porter and Moss, Topographical Bibliography IV, p. 148.

66 This diagram is derived from photographs taken by Judy Lustig combined with the divisions indicated by Newberry in BH I, Pls. 15 and 16.

67 This assumption is made on the basis of the titles of this man found both in his tomb and in this inscription, and on the connection of Khnumhotep I with Amenemhēt I evidenced in his autobiography in Tomb 14.

68 Lines 25-46.

69 Lines 46-53.

70 Lines 54-62. This biography only mentions the title Overseer of the Eastern Desert for Nakht, but in his own tomb (21), this man lists Great Chief of the Oryx Nome as one of his titles.

71 Lines 71-79.

mother, Bakt. His father was a count named Neheri, whose birthplace is not given.[72] Khnumhotep II was married to Khety, a princess of the neighboring Jackal nome (U.E. 17); their eldest son, Nakht, was appointed ruler of his mother's nome by Senwosret II.[73]

The remainder of the inscription tells of Khnumhotep II's building projects in Menat Khufu, including the erection of ka chapels (*ḥwt-k3*) and the setting up of statues, possibly to honor his ancestors,[74] but also probably for himself.[75] He emphasizes his dutifulness and devotion in restoring the name of his family on monuments which had fallen into disrepair, and describes the building of his tomb, comparing his actions to those of his father, who had built a ka chapel in the town of Mernofret.[76] The remainder of the autobiography emphasizes the close relationship of his family to the royal house and Khnumhotep II's generally excellent character, and asks that offerings be made to him regularly.

BURIAL SHAFTS

Two large rectangular burial shafts were cut into the floor along the south wall of this chamber, one in the center of the wall and the other at the west end. A third shaft was marked out on the floor at the east end of the south wall, and a fourth, probably later in date, was begun in front of the doorway to the shrine. When the central shaft was cleared, it was found to descend vertically for 6.6 meters, and then step down roughly for another 2.8 meters to the burial chamber, which consists of a central passage with two niches along the sides and a third niche at the far end. Newberry suggests that the side niches were each meant to hold a sarcophagus, while the far chamber was cut to hold a foundation deposit and the sarcophagus of Khnumhotep II himself.[77] Both burial shafts had been plundered in antiquity; the wells contained 12th Dynasty, Coptic, and late Roman pottery. Fragments of sarcophagi painted in blue, red, black, and yellow were also found, along with a rough offering table, the top of a small limestone obelisk which may have

---

[72] Lines 68-69. Newberry suggests that Neheri was connected with the 15th U.E. nome, the Hare nome, which borders the Oryx nome to the south (cf. Newberry, BH II, p. 9). However, Redford notes that the town of Mernofret, where Neheri is said to have built a ka chapel (see below), can be identified with a tomb in the area of Tihna, which he locates in the 18th U.E. nome, contiguous to the Oryx nome on the east bank. He therefore suggests that Neheri comes from this nome. (Redford, "The Father of Khnumhotpe II," pp. 158-59.) For the location of Mernofret, see also Kessler, Topographie, pp. 258ff.

[73] Lines 121-50.

[74] Breasted, Ancient Records II, p. 285, note a.

[75] Lines 80-99.

[76] Thought to be located where the modern town of Tihna is today (Kessler, Topographie, pp. 258ff). See also Redford, "The Father of Khnumhotpe II."

[77] Newberry, BH II, p. 73.

stood at one side of the offering table, parts of a necklace of blue glaze and several beads, and three skulls.[78]

SHRINE

A doorway in the center of the east wall of the main chamber, approached over a raised threshold, opens into a small rectangular shrine, approximately 2.5 meters wide and 3 meters deep. The floor of this shrine is on the same level as the main chamber, but the ceiling is considerably lower. A portion of a rock-cut throne projects from the east (back) wall, indicating that there was once a large seated figure of the tomb owner here;[79] a portion of the face of this statue was found outside the tomb, on the scree slope.[80] Flanking what remains of Khnumhotep II's throne are standing figures, incised and painted, of three women. On the left is Khety, Khnumhotep II's wife; to the right, one of the women is his mother, Bakt. The label associated with the third woman, who is to the right of Bakt, is completely destroyed; therefore, she cannot be identified with any certainty.[81] Perhaps she can be identified with Khnumhotep II's maternal grandmother, Satip.[82]

ARCHITECTURAL MODELS

The architecture of this tomb, as mentioned above,[83] corresponds to both the elite house and the god's and royal mortuary temples of the New Kingdom.[84] This circumstance is not surprising, since both the tomb and the temple have been shown to be modeled on the elite house, and thus must logically bear some resemblance to one another.[85] In fact, however, the correspondence between this tomb and later Egyptian temples is even stronger than the resemblance between either the tomb or the temple and the Egyptian house.[86]

New Kingdom temples can be extremely elaborate structures, with multiple shrines and large numbers of magazines and other rooms added to the central core. However, the basic form of the temple is relatively consistent. The Khonsu Temple at Karnak is used for comparative purposes here, since it is fairly simple and quite well preserved. The form of this temple reflects the key elements found in most temples of the New Kingdom and later.

---

78 Newberry, BH II, pp. 73 and 79. The excavator decided that one of the three skulls, "...remarkable for the general massive appearance and firm square jaw..." could perhaps have been Khnumhotep II himself; he therefore re-buried it in the central shaft.

79 Newberry, BH I, p. 53.

80 Newberry, BH II, p. 72.

81 Newberry, BH I, pp. 71-72.

82 See below, microscene Sh1.

83 See Chapter I.B.

84 This type of temple is referred to as Mature Formal (Kemp, Anatomy, p. 185).

85 Cf. Gardiner, Attitude, p. 10.

86 Cf. Badawy, Architecture II, p. 195.

The temple is entered through a pylon, which leads to an open courtyard. Tomb 3 probably had no pylons, but the open courtyard which lies in front of the entrance clearly corresponds to the open courtyard of the temple. Beyond the courtyard is a large columned, or hypostyle, hall. This hall corresponds to the main chambers of the tomb chapel. Behind the pillared hall of the temple is a sanctuary, which would presumably have contained one or more cult images, corresponding to the shrine of Tomb 3, which contains images of the tomb owner and three women. A closer parallel to Khnumhotep II's shrine can be found in rock cut temples, such as the temple of Ramesses II at Abu Simbel: the shrine at the back of this temple contains four rock-cut statues.[87] The plans of any of a number of temples can be shown to correspond to the plan of Khnumhotep II's chapel; a plan of the Ramesside temple of Beit el-Wali is also included here since it, like the tomb chapel, is rock-cut, and contains only the basic elements of the temple.

Figure II.6: Comparison of Beit el-Wali (Ramesside), Tomb 3, and Khonsu Temple, Karnak (Ramesside)

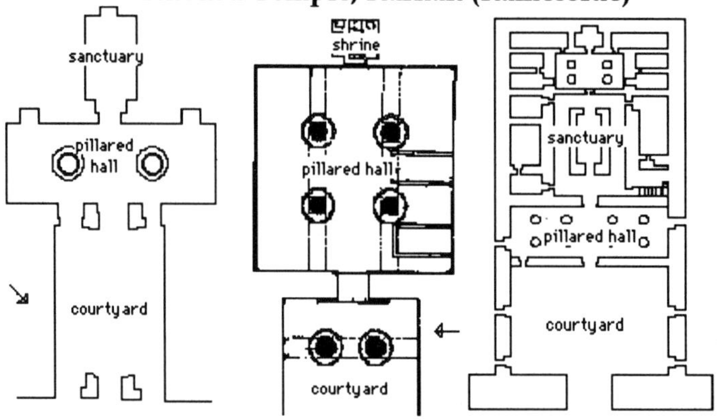

Although the layout of the New Kingdom temple is somewhat more elaborate than that of the chapel of Tomb 3, the basic elements correspond. In both cases, the buildings are organized along a central axis, with each successive room opening directly from the one before in a straight line. When temples contain additional subsidiary rooms, they are placed off this central axis.[88] In both cases, and indeed, in most temples, this is an east-west (or west-east, depending on its location) axis, which can be associated with the path of the sun.[89]

---

87 Otto, "Abu Simbel," LÄ I, col. 26.

88 Cf. Gundlach, "Tempelrelief," LÄ VI, cols. 408-409.

89 Vittman, "Orientierung," LÄ IV, cols. 607-608.

Figure II.7: The Ka Chapel of Heqaib[90]

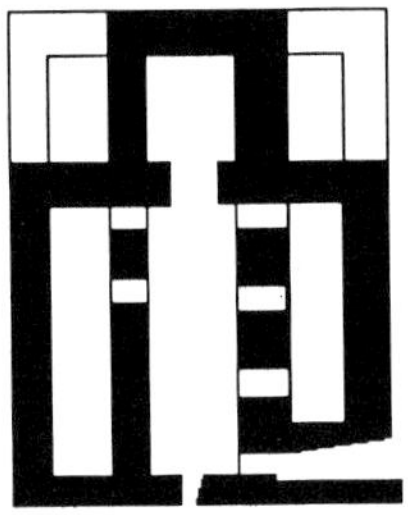

A very close architectural correspondence to this tomb is also offered by the 11th Dynasty ka chapel of Heqaib at Elephantine, demonstrating that this type of plan was in use in cultic structures other than tombs before the 12th Dynasty. The excavator's reconstruction of this structure shows a square main chamber with two longitudinal rows of heavy brick piers, and a smaller shrine area. The piers divided the principal chamber into three aisles, each of which was roofed with a vault. The outlines of this building are thus very similar to the chapel of Tomb 3, although less elegant and spacious.[91]

---

90 After Habachi, The Sanctuary of Heqaib, p. 156, figure 10.

91 Habachi, The Sanctuary of Heqaib, pp. 156-57 and note 40. It should be noted here that the chapel reconstructed here was replaced in the early 12th Dynasty by a more complex structure.

Egyptian temples of the New Kingdom is used as comparative and supportive material.[28]

A number of scholars have addressed the issue of the overall function of elite tomb decoration. One older synthetic treatment of a specific Egyptian tomb is to be found in Gardiner's commentary in Davies' 1915 publication of the 18th Dynasty tomb of Amenemhēt at Thebes. In this work, Gardiner notes that, at least for Theban tombs of this period, "not only is the choice of the pictures closely dependent on the contemporary conception of the meaning of the tomb, but also their exact position within it is fixed in accordance with a keen sense of situation and orientation." Thus, for example, he states that the reliefs in the outer rooms express the concept of the tomb as the house of the deceased, and are decorated with biographical texts and depictions of worldly activity, the long passage illustrates the funeral, and the inner rooms portray life in the Hereafter.[29]

Gardiner sees the scenes and texts in this tomb as reminders of the past life of the deceased on earth, which magically provide for similar activities and amusements in the Afterlife, and also serve to enhance the reputation of the tomb owner. He even suggests that the decoration of houses of the noble class was similar to that seen in tombs, rendering the identification of chapel with house even more relevant; however he provides little evidence for this supposition. He then considers the burial chamber to be part of the netherworld, conceived of as "an intricate labyrinth of interconnecting halls and portals." Gardiner also points out the importance of the orientation of important elements, such as the convention that the deceased is shown facing out, and visitors are depicted with their backs to the entrance of the tomb.[30] Since Gardiner is dealing with a New Kingdom tomb, his conclusions are not necessarily applicable to the tomb being studied in this work, but he raises interesting issues and at least attempts a unified approach.

More recently (1983), Hartwig Altenmüller[31] has looked at the development of the titulary, biography, and a number of types of tomb scenes over the course of the Old Kingdom, and has concluded that the function of these elements changes subtly over time. The gradually changing function of the decoration must, according to this scholar, be seen against a backdrop of the most favorable portrayal of the tomb owner possible, which was designed to assure him or her of immortality, and of an appropriately high status in the Afterlife.[32]

---

28 See above, Chapter V.D.

29 Davies and Gardiner, Amenemhēt, pp. 17-19.

30 Davies and Gardiner, Amenemhēt, pp. 19-20. On this subject, see also Kanawati, in "Living and dead," pp. 219-21, and Arnold, Wandrelief, p. 128.

31 "Lebenszeit und Unsterblichkeit in den Darstellungen der Gräber des Alten Reichs."

32 Altenmüller, "Lebenszeit," p. 87.

Altenmüller traces this concept from the early Old Kingdom, when significant amounts of decoration, namely titulary, limited figural representation, and the beginnings of the tomb biography appeared as an integral part of the tomb. According to his theory, these elements were originally designed to express the relationship of the deceased to the king. In the mid-4th Dynasty, along with the professional life of the tomb owner, which was directly dependent on the king, the deceased's personal life in all its phases appears to be gaining in importance.[33]

The 5th Dynasty saw an increase in the repertoire of tomb decoration, including scenes showing the tomb owner in connection with official duties with which he should, by virtue of his titles, have had no direct involvement.[34] Altenmüller attributes these to the issue of provisioning, both of the individual funerary cult, and of the larger society of the Egyptian state. Emphasis is now on service to the sun god, as guarantor of the maintenance and care of humankind on earth, rather than to the king.[35] By the end of the Old Kingdom, the emphasis in tomb decoration was on the service of the deceased to humankind in general, as expressed in the "ideal biography" of the 6th Dynasty. These biographies, which stress the positive deeds of the tomb owner and ethical behavior toward social inferiors, serve to underline the care that the deceased has taken of his or her social inferiors, and thus of humankind as a whole.[36]

A recent work by Claudio Barocas (1982)[37] investigates the function of decoration within elite tombs, specifically Memphite tombs of the 4th-5th Dynasties. This scholar approaches his data with the desire to create a link between the architectural purpose of the tomb structure in general, which he sees as clearly funerary, and the scenes found decorating some of the tomb chapels, which he does not consider obviously mortuary in character.[38] He finds this link in the offering cult, and sees tomb decoration as an extension of the desire for offerings. His primary conclusion seems to be that the tomb is built and decorated to pacify the dead so that they do not come back to haunt

---

33 Altenmüller, "Lebenszeit," pp. 76-78.

34 This is supposedly in constrast with scenes from the late 3rd and very early 4th Dynasties, for example, in the tomb of Metjen, where the tomb owner, the master of the hunt, oversees hunting (Altenmüller, "Lebenszeit," p. 78). However, other tombs from this period, which do not belong to appropriately titled owners, also depict this type of scene (for example, the tomb of Rahotep and Nofret, see Smith, HESPOK, pp. 153f.).

35 Altenmüller, "Lebenszeit," p. 79. He also notes that these types of scenes are based in reality: when they first appear, they are connected with the appropriate officials.

36 Altenmüller, "Lebenszeit," pp. 83-84.

37 "La Décoration des chapelles funéraires égyptiennes."

38 Barocas, "Décoration," p. 429. In fact, however, he seems to label almost all of the scenes "daily life" scenes, regardless of their character. (Barocas, "Décoration," p. 430.)

the living. The main impetus behind the decoration of the Egyptian tomb would thus be fear of the dead.[39]

Like Kessler and Weeks, Barocas notes that the categories we often assign to various scenes (such as "daily life") were not necessarily those recognized by the ancient Egyptians, and emphasizes that scenes with which the tomb owner is associated often show activities for which he or she probably would not have been responsible in life. He offers alternative interpretations of these categories as follows. Scenes of "work," that is agriculture, food production, manufacture, etc., which are performed by others under the supervision of the deceased, are carried out with a view to keeping the deceased content, by emphasizing the successful completion of certain tasks which were necessary for the offering cult. Scenes of funerary ritual are depicted with the goal of proving that burial rites have been accomplished properly. He interprets the third scene type, epitomized by depictions of the tomb owner hunting in the marshes, as images of the deceased single-handedly gathering food for his family, and concludes that such scenes qualify the noble in the eyes of his family as worthy of receiving offerings.[40]

Barocas suggests that all of these scenes were developed as an extension of the offering rite, which he sees as paramount in the decoration of the early 4th Dynasty. He also suggests that the offering itself served primarily to placate the deceased; thus the decoration, as an extension of the offering, also served to pacify the deceased. A reciprocal relationship is demonstrated: the living work to assure the funerary offerings and rituals (types 1 and 2) because they recognize the place which the deceased occupied in society, and the importance of his or her activities (type 3).[41] According to this theory, the concept of tomb decoration has more to do with the relationship between the living and the dead than it does with the desire of the tomb owner for a good Afterlife, although this latter element is still present.[42] Thus Barocas arrives at the conclusion that the tomb superstructure, with its decorated chapel, was designed to keep the deceased from coming back to earth by providing rituals that would satisfy him or her and remove any possible reasons that he or she might have for returning to the world of the living.[43]

There are a number of interesting ideas contained within this treatment, and, in fact, some of it works well with the Beni Hasan data, but the validity of the central hypothesis is not shown convincingly within the framework of this short article. Although Barocas's ideas are interesting and potentially valid,

---

[39] Barocas, "Décoration," pp. 430-31. This is an interesting point, but one for which Barocas does not present convincing evidence.

[40] Barocas, "Décoration," pp. 434-45.

[41] Barocas, "Décoration," pp. 433-35.

[42] Barocas, "Décoration," pp. 430-33.

[43] Barocas, "Décoration," p. 436. Again, there is little evidence presented here; even fairly obvious sources such as the Letters to the Dead have not been cited.

they are not fully supported with hard data, perhaps due to the difficulties of dealing with a large corpus of tombs.

Finally, Yvonne Harpur (1987)[44] studies the organization of a number of scene types in the entire corpus of Old Kingdom tombs. In this work, Harpur traces the various scenes that are found in different parts of tombs from different dynasties and different sites, tracking details such as figure orientation. This work is primarily a chronological and locational study, designed to provide a way of proveniencing relief fragments not found *in situ*, therefore its application to the study at hand is limited. Harpur's graphic presentations, however, are quite useful, and have been used as partial models for many of the diagrams presented in this volume.

Several recent dissertations also address issues of tomb decoration in both cases, for New Kingdom tombs. Meyers has studied several of the Amarna tombs, and suggested ways in which political ideology is reflected by the decoration.[45] Shaheen has investigated selected occurrences of tribute scenes in 18th Dynasty tombs.[46] Both works have been useful in suggesting aspects of the approach and methodology used here.

---

44 Harpur, Decoration in Egyptian Tombs of the Old Kingdom, London and New York; 1987.

45 Meyers, A Program of Political Theology in Amarna Tomb Art, University of Pennsylvania; 1981.

46 Shaheen, Historical Significance of Selected Scenes involving Western Asiatics and Nubians in the Private Theban tombs of the XVIIIth Dynasty, University of Pennsylvania; 1988.

CHAPTER IV : DECORATION OF TOMB 3

The tomb chapel of Khnumhotep II contains 44 microscenes: 16 on the west wall, 11 on the north wall, 5 on the east wall, 7 on the south wall, and 5 in the shrine. This chapter presents descriptions and suggested interpretations of each of these microscenes and overviews of their immediate spatial contexts (the four walls of the main chamber, and the shrine as a whole).

WEST WALL

The majority of the scenes on this wall are concerned with the manufacture of non-perishable cult or burial equipment and the production and preparation of food. The manufacturing of non-perishable items and the processing of food is shown to the south (left) of the entrance doorway, while the growing, collecting, and storing of various foodstuffs are to the north (right).[1] Also on this wall is a statue transport scene and riverine voyages to and from Abydos. This entire wall is illustrated in Plate 1.

---

[1] Left and right are given according to the viewpoint of an observer facing the wall.

Figure IV.1: West Wall, Microscenes

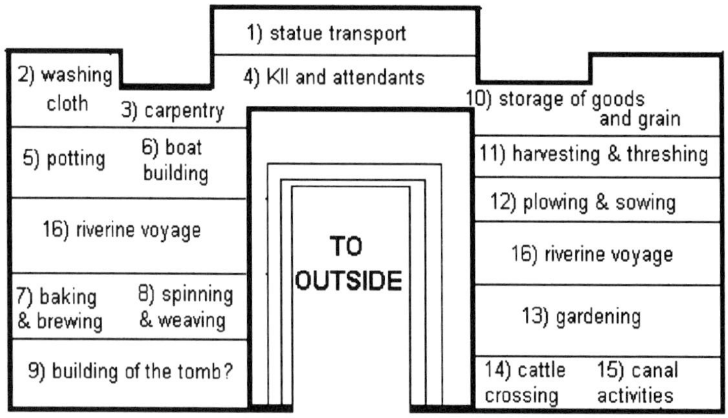

W1. Statue Transport

This scene occupies the whole of the uppermost register above the doorway into the tomb. The activity on the right appears to center on a shrine, its two doors open to reveal the depiction of a standing male statue, shown in true profile. This statue wears a knee-length starched and pleated kilt and long wig, and holds a staff in its right hand. There are several indications that this statue was meant to be of Khnumhotep II himself. First, in Old and Middle Kingdom tombs, labeled images in statue transport scenes such as this one always belong to the tomb owner.[2] In fact, a similar scene, pre-dating this one by about a generation, appears in Tomb 2 at Beni Hasan; the statue here is specifically identified as an image of the tomb's owner, Ameni.[3] Second, decorum at this point in time seems generally to have precluded images of gods or kings from appearing in private tombs,[4] and this statue is not labeled as Khnumhotep II's father or any other of his ancestors. This image also resembles representations of Khnumhotep II which appear elsewhere in this tomb, especially the large-scale figure on the north wall (also bearded, wearing a long wig, and holding a staff).

Above the shrine is the label: "Escorting the statue to the temple,"[5] a phrase which suggests that this statue is intended to be set up in the context of a local temple, perhaps as part of a permanent cult. In his biography,

---

2 See Eaton-Krauss, Statuary, Chapter 5.

3 See Newberry, BH I, Pl. 13.

4 Sorenson, "Divine Access," pp. 112-13. One exception to this is found in the tomb of Intefoker (see Davies, Antefoker, Pl. 16).

5 *šms tw(t) r ḥwt-nṯr*. Cf. Eaton-Krauss, Statuary, p. 64, and Silverman, " The Nature of Egyptian Kingship."

Khnumhotep II mentions setting up statue cults: "...I followed my statues to the temple, and I put in order for them their offerings of bread, beer..."[6]

Figure IV.2: Statue Transport

The existence of both royal and private statue shrines (*ḥwt-k3w*) associated with divine temples is archaeologically attested from the Old Kingdom and later. At Abydos, for example, there is a complex of such structures spanning the Old, Middle, and New Kingdoms, which was presumably connected with the nearby temple of Khentyimentiu. Little remains of the Middle Kingdom stratum, but the Old Kingdom examples, although mostly royal, provide some evidence for the ownership of some of these chapels by private people, albeit through royal favor.[7] Evidence for similar shrines has been found at Hierakonpolis, Bubastis (all royal), and Elephantine (one or perhaps two private), and one royal ka chapel belonging to King Nebhepetre Mentuhotep was excavated at Dendereh.[8]

The practice of setting up private statue cults in temples is also attested textually in both the Old and Middle Kingdoms.[9] A famous example of such a reference comes from the Middle Kingdom tomb of the nomarch Djefaihapi at Assiut,[10] who had contracts detailing the upkeep of his statue cult(s) in local temples inscribed on the east wall of his tomb. These contracts detail the endowment and outline the process by which certain offerings to these local temples were to be passed on to his statue cults.

---

6 Lines 83-84: *...šms.n.i twwt.i r ḥwt-nṯr sm3.n.i n.sn p3t.sn t ḥnḳt. ..* (Cf. Eaton-Krauss, Statuary, p. 191 and note 940.) This description appears after a section in which Khnumhotep II discusses perpetuating the name of his father and "endowing the ka temples appertaining thereto" (lines 78-83), thus it is conceivable that he is referring here to statues of his ancestors, as is suggested by Breasted (Ancient Records, p. 285). He calls these statues *twwt.i*, "my statues," thus either these statues were offered by him for his ancestors, or offered by him for himself. It is most likely, however, that the statue here, as discussed above, is of Khnumhotep II himself.

7 O'Connor, "The Status of Early Egyptian Temples," p. 8.

8 See O'Connor, "The Status of Early Egyptian Temples." For the last, see O'Connor, "Nephepetre's Chapel."

9 Helck, "Opferstiftung," LÄ IV, cols. 592-594.

10 See Breasted, Ancient Records I, pp. 260ff., for translation, and Reisner, "Hepzefa," and Spalinger, "Redistributive Pattern," for translation and analysis.

The Djefaihapi inscriptions also describe the movement of one or more statues to and from the main temples of the Assiut area. Reisner hypothesized that there was only one statue involved; this statue originally, according to his reconstruction, would have been set up in a valley temple associated with Djefaihapi's tomb and then taken in procession to the temples of Wepwawet and Anubis during certain festivals.[11] Reisner's analysis of these contracts provides, therefore, another possible explanation of this scene, i.e. that it represents a statue procession that is taking place in the context of a festival. This theory is interesting in light of the proximity of the festival list that is located on the exterior face of the west wall, on the lintel over the entrance to the tomb,[12] and is thus spatially connected with this scene. Support for an identification of this scene as a festival procession can also be found in Khnumhotep II's biography, which lists festivals at which invocation offerings were to be made to his statues.[13]

Behind the shrine are two officials: "the Overseer of Sealers, Ima(t),"[14] who also appears in microscene W10, and a man named Khnumhotep. This latter may be the tomb owner himself, escorting his own statue during life, but it is more likely that this figure represents one of Khnumhotep II's sons, two of whom were also named Khnumhotep.[15]

Directly in front of the shrine is a priest who offers incense to the statue. To his left is a lector-priest who holds a papyrus. Above this lector priest is written his title and the label: "*Ḥr mꜣꜥ ḫrw.*" This phrase appears twice again on this wall, again over the figure of a *setem*-priest in the context of the pilgrimage to Abydos (microscene W16) and over a man who kneels inside the depiction of a shrine in microscene W9; it also appears on the south wall of the main chamber, in association with a lector priest, and the shrine of this chapel, also above the figure of a lector priest.

Newberry translates this phrase as a name, "Hor-maacheru," as does Ranke.[16] However, this is unlike any other name found at Beni Hasan. If it were a name, one would have to assume either that there are at least two, if not three, distinct people who bore it: the lector-priest here and another in the shrine, a *setem*-priest on the funerary boat, and an untitled official inside the shrine in microscene W9, or that one person played all of these different roles.

---

[11] Reisner, "Hep'zefa," pp. 90-91.

[12] See Newberry, BH I, Pl. 29.

[13] Lines 89-95: *prt-ḫrw m ḥb nb n ẖrt-nṯr m tpy rnpt ? rnpt ꜥꜣ rnpt wrt ꜥrḳ rnpt ḥb wr m rkḥ ꜥꜣ m rkḥ nḏs m* 5 *ḥrw rnpt m šḏtꜥ šꜥ m ꜣbd* 12 *ꜣbdt* 12 *ḥb nb n tp tꜣ nfr ḏw.*

[14] *imy-r sḏꜣwty imꜣ.*

[15] Eaton-Krauss (Statuary, p. 192 and note 940) suggests that the hieroglyphs here, which I read as two separate labels, belong to one inscription, to be read *sḏꜣwty-bjtj [ꜥ-]jꜣm H̱nm-ḥtp;* she suggests also that this man is one of Khnumhotep II's sons.

[16] Newberry, BH I, p. 51; Ranke, Personennamen, 144.22; II, 359.

Although it is possible that this formulation represents a name, it seems more likely that *Ḥr m3ʿ ḫrw* may represent a label or a phrase that was recited.[17] The expression *m3ʿ ḫrw*, translated "true of voice," "justified," or "acclaimed as right," is thought to refer originally to the outcome of a mythical judicial proceeding, modeled after trials between Horus and Seth. It appears first in the context of the Pyramid Texts, where it refers to the king as Horus, who is legitimized by this phrase as heir to Osiris and thus king of Egypt;[18] it appears also in the Coffin Texts, where it refers to the deceased person.[19] From the Middle Kingdom on, it is used more universally, in the phrase "N is justified," or "N, justified," and is usually interpreted as identifying the deceased person with Osiris.[20] This Osirian identification is indeed strong in the later periods of Egyptian history, but in the early Middle Kingdom, identification with Horus may have also been implied by this phrase, as identification of the deceased with Horus is certainly made explicit in the Coffin Texts.[21]

This may well be the case in regard to the labels discussed here, since it is the original phrase, "Horus is justified!" with which we are dealing. There are several possible interpretations of this phrase. One is that it is recited by the various priests and, like the comparable phrases in the Pyramid Texts and Coffin Texts, identifies the deceased (here Khnumhotep II) with Horus, and thus legitimizes him as the heir of Osiris. It is also possible that the justified Horus refers to the eldest son, Nakht, who is the chief celebrant of his father's cult, and would, as Horus, be responsible for the burial of his father as Osiris. It seems somewhat more likely that Khnumhotep II himself is being proclaimed a justified Horus by the lector-priest, since this phrase and the other examples of this label within the chapel appear to be associated with images of the tomb owner; however, both possibilities must be considered.

Another phrase is inscribed above the crowd of men who drag the shrine: "the god will come that the earth might be protected."[22] This phrase appears in each of the statue transport scenes at Beni Hasan.[23] The word *nṯr*

---

17 The text faces away from the lector-priest; it would be more usual for the text to face him if he were reciting it (cf. Fischer, Orientation, pp. 48ff.).

18 Pyr. 2089a in Utterance 689; see Anthes, "The Original Meaning of M3ʿ-Ḫrw," pp. 43 and 48ff.

19 See references in Grieshammer, Das Jenseitsgericht, pp. 152-57.

20 Brunner, "Maʿa-cheru," LÄ III, col. 1108.

21 Grieshammer, Das Jenseitsgericht, pp. 19 and 43-45. See for example, Faulkner, The Coffin Texts, Spell 1, p. 1, and Spell 7, pp. 3-4.

22 *ii nṯr s3 t3*. Cf. Lüddeckens, who translates this phrase in New Kingdom contexts as an imperative: "Come, god, in order to protect the earth," ("Totenklagen," pp. 52 and 62). See also Silverman, "Kingship in Literature."

23 In addition to this scene in Tomb 3, such scenes are found in Tombs 15, 17, and 2 (see Newberry, BH II, Pls 7, 13, and 17, and BH I, Pl. 13).

can be translated as "god,"[24] thus the inscription appears to refer to the bringing of a divine image whose function is to protect the earth and therefore to play a role in maintaining the cosmic order.

A second crowded group of men faces the shrine and its transporters. The first figure in this group is labeled: "the Wab-priest, Khety;"[25] he, and at least one of the other men hold their arms out as if clapping or singing. Above this group is the inscription: "opening of the doors of heaven that the god may come forth."[26] In the other Beni Hasan tombs containing this scene, the statue is always met by groups of musicians; these men, led by a priest, may be fulfilling this musical role.[27] Phrases similar to this one are also found in the context of later royal and divine inscriptions. An almost identical inscription comes from a text describing the solar apotheosis of King Amenhotep III: "Opened are the doors of heaven so that the god may come forth pure."[28] The inscription in Tomb 3 evidently refers at one level to the doors of the shrine, which are opened so that the divine image of Khnumhotep II can emerge (and be purified by the accompanying Wab-priest). It is also likely that shrines in general could have been identified as "heaven," since they served to link the earthly and celestial realms by providing dwellings for divine beings.[29]

The final group of figures in this scene consists of five female dancers, each of whom wears a pair of short dark briefs and has her hair dressed in a club on the top of her head; this style is reminiscent of the headdresses of the *Mww* dancers who often accompany or meet the funeral procession in Old Kingdom tombs.[30] These women form two distinct vignettes. At the left, two of the dancers make a "smiting the enemy" tableau, with the first woman facing right, grasping her partner's headdress/hair with her left hand and raising her right arm above her head. Above these dancers is written: "Under (my) feet."[31]

---

24 See WB II, p. 358. According to the references contained within this citation, this term is used to refer to the divinized dead in the Pyramid Texts, Coffin Texts, and Book of the Dead.

25 *wꜥb Ḫty.*

26 *wn ꜥꜣwy pt pr nṯr.* Cf. Brovarski, "Doors of Heaven," pp. 107-110; Wente, "Hathor at the Jubilee," p. 88.

27 See for example Tomb 2 (Newberry, BH I, Pl. 13).

28 See Wente, "Funerary Beliefs," p. 24; Wente, "Hathor at the Jubilee," p. 88.

29 Černý, in "Note on *ꜥꜣwy pt* 'Shrine'," suggests that in the New Kingdom at least, the phrase *ꜥꜣwy pt* can be translated as "shrine." Brovarski suggests that a structure occasionally seen in Old Kingdom representations of the funeral cortege, and apparently representing a purification booth set up on the terrace of the royal valley temple, can be identified as the "doors of heaven." (See Brovarski, "Doors of Heaven," pp. 107-113.)

30 The *mww* dancers usually appear in one of six contexts: in the "Hall of the *mww*"; at the Journey to the West; at the Journey to Sais; at the dragging of the sarcophagus; at the procession of the tekenu and the canopic jars; and at the Tekenu-ritual. (Altenmüller, "*Mww*," p. 3.) The same or a similar hairstyle is also worn by one of the female bakers in microscene W8.

31 *ḫr rdwy.(i)*

This dance may pantomime the victory of the king over his enemies,[32] and if so suggests an identification of Khnumhotep II with the king, or perhaps even with Osiris as the newly installed king of the underworld, whose enemies are overcome by Horus.

The second dance involves three women, two of whom bend backwards under the raised arm of the third. Brunner-Traut suggests that the standing woman in this vignette is the "wind,"[33] before whom the other two women bend like grass.[34] This could also be some sort of victory dance. It has also been suggested that these women are acting out the "Song of the Four Winds," a version of which is preserved in chapter 162 of the Coffin Texts.[35] This spell alludes to the power which the deceased holds over these winds, which came into being "before men were born or the gods existed," and provide the "breath of life."[36] Dance in general is often associated, both in the Middle and New Kingdoms, with the worship of Hathor, and may have overtones of themes of regeneration.[37]

The statue transport scene as a whole can therefore be considered to encapsulate the installation of a statue of Khnumhotep II at a local temple, or possibly even of a series of statues at local temples. (Such temples are not named here, but could have been dedicated to Khnum of Herur or one of the other gods mentioned within the tomb.)[38] At the same time, this scene represents subsequent festival processions during which such a statue or statues would have moved about, perhaps from temple to temple, or from tomb to temple. Statues of deceased persons were thought to provide repositories for their kas, a term which can be defined as the "vital force" of a human being that was present at birth and during life, but which was separated from the body at death and reunited with its owner through the proper celebration of the funerary ritual.[39] This statue cult may have been represented in the tomb to associate it with the funerary cult and thus reiterate its importance and ensure its upkeep, just as Djefaihapi inscribed the contracts for his statue cult on the walls of his tomb. The statue, and therefore the tomb owner, is clearly granted divine status here, and may even be identified specifically with Horus, Osiris, and/or the king.

The scene also has geographical significance, since it refers to the location of one or more local temples, which would most likely have been built in the western floodplain, where the major towns of the area are thought to

---

32 Brunner-Traut, Tanz, p. 37.

33 *ṯ3w*.

34 Brunner-Traut, Tanz, p. 37.

35 Vandier, Manuel, IV, pp. 426-427; see also Brunner-Traut, "Tanz," LÄ VI, col. 225 note 1.

36 For translation, see Faulkner, Coffin Texts, I, p. 140.

37 Brunner-Traut, Tanz, pp. 39-40; Wente, "Hathor at the Jubilee," pp. 88-89.

38 See Table 2.2.

39 Spencer, Death in Ancient Egypt, pp. 58-59.

The figure of Khnumhotep II which watches the boat building is seated in a carrying chair borne by four men, who are shown in a fairly realistic manner, overlapping so that only the man nearest the viewer is completely visible. The arms of the chair are decorated with the hieroglyphs for life, wealth, and stability, and the base is ornamented with papyrus clumps. Khnumhotep is only visible from the chest up, but he once again appears to be wearing an enveloping cloak similar to the one he wears in the register above (see microscene W4); again, this probably indicates that he is out of doors. Behind him walks an attendant who holds a hide-covered sunshade[63] over him with one hand and carries an axe in the other. The sunshade also indicates that the activity would have taken place outdoors.

Figure IV.7: Boat Manufacture

The placement of this scene adjacent to a river voyage, as well as the content of the scene itself, suggests a location on the riverbank. The adjacent potting scene reinforces this hypothesis. It has been suggested that wood was a royal monopoly, and that shipbuilding was therefore a province of the central government, with shipyards located at the royal court.[64] However, the presence of Khnumhotep II himself in this scene, and the implied use of acacia, which grew all over Egypt,[65] implies that boats would have been manufactured on the tomb owner's estate.

The label above this figure of the tomb owner reads: "Inspecting the shipbuilding by the Hereditary Prince, Count, Khnumhotep."[66] Kessler notes (but unfortunately does not document) that the building of a ship is a task undertaken by Horus during the New Year's festival.[67] This theme does appear in the "Contendings of Horus and Seth," where Horus makes a boat of

textual evidence, to have been the principal building material for ancient Egyptian wooden boats. (Vandier, Manuel V, pp. 680-81, 683; Klebs, Reliefs II, p. 137.)

63 According to Newberry's copy, this sunshade is depicted in a perspective view; cf. Schäfer, Principles, p. 265.

64 Kessler, "Szenen des Schiffsbaues," p. 67.

65 Cf. Helck, "Akazie," in LÄ I, col. 113, for use of acacia in shipbuilding.

66 *m33 mḏḥ šw n r-pꜥ ḥ3ty-ꜥ Ḫnm-ḥtp.*

67 Kessler, "Szenen des Schiffsbaues," p. 67.

wood, and Seth a boat of stone,[68] but a link between such myths and this scene is not easily supported.

The few inscriptions associated with shipbuilding scenes, which date primarily to the Old Kingdom, indicate that the ships being built were connected with the funeral.[69] There is no such inscription in Tomb 3, but the close proximity of the shipbuilding to the pilgrimage scenes (W16) suggests that this boat, and the others implied by the scene, would in fact have been used for the voyages to and from Abydos, a journey connected with, if not also part of, the funeral.[70]

W7. Brewing and Baking

The preparation of various types of food, which occupies a great deal of space in Tombs 15 and 2, is limited to one partial register here, and only brewing and baking are depicted. The vignettes comprising this scene are arranged in seemingly random order, as if to represent more realistically the interior of a bakery-brewery such as is illustrated by a wooden model found in the tomb of Meketre.[71] Kitchens would presumably have been located on the estate of the tomb owner, probably within the walls of the house area proper.[72] Both bread and beer were made from grain,[73] the planting and harvesting of which is shown across the doorway in microscenes W11 and W12.

Figure IV.8: Brewing and Baking

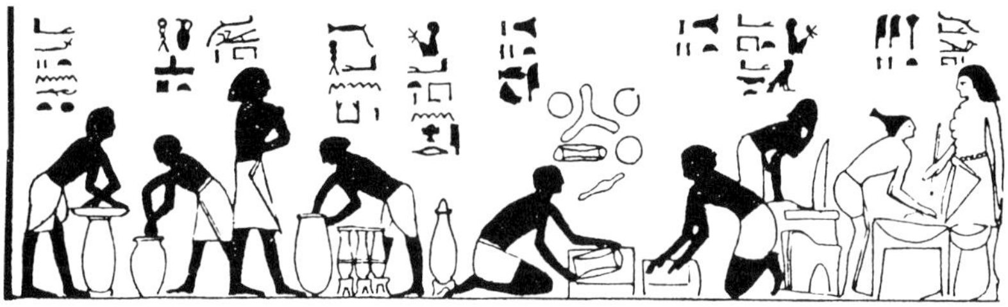

The brewing is to the left, supervised by an overseer of a storeroom; his workers strain the beer and put it into jars.[74] Above the vignette of a man putting beer from a vat into smaller jars is a label which reads: "filling for the

---

68 See Lichtheim, Literature I, pp. 220-21.

69 Kessler, "Szenen des Schiffsbaues," p. 67.

70 See below, microscene W16.

71 Winlock, Models, Pls. 22-24, 64-65; model G, pp. 27-28. See also Kemp, Anatomy, pp. 120ff.

72 See Kemp, Anatomy, pp. 296-97.

73 See Wilson, Food and Drink, pp. 11-20.

74 Cf. Helck, "Bier," LÄ I, col. 789-92.

ka."[75] This leaves little doubt as to the eventual use to which the beer will be put: it will be offered in the funeral and the funerary cult of the tomb owner. In fact, beer is seen on the south wall, carried by offering bearers, on the offering tables and on the menu, and also appears in the shrine.

The rest of the figures in this scene are involved in bread-making. The grinding of grain, the shaping of loaves and cakes, and the cooking of these loaves in an oven are shown. The grain is ground by a woman, who has her hair clubbed on top of her head in the same hairstyle worn by the dancers in the statue transport scene (microscene W1). "The Guard of the Chamber of Bakers, Meket"[76] opens the lid of an oven to check its contents.

Bread and beer were the most important foods in the Egyptian diet; the minimum food needed for existence was thought to be five loaves of bread and two jars of beer per day.[77] Bread and beer appear in all funerary menus[78] and on offering tables, and are used as determinatives in the invocation offering formula.[79] Thus by depicting their manufacture, even in scenes as abbreviated as these, their perpetual supply was ensured. Both items are featured prominently in Khnumhotep II's and Khety's offering cults, appearing on the south wall, in the shrine, and on the offering menus. It seems likely that the bread and beer making here stands generally for the processing of food which can be seen in greater variety in other of the tombs at Beni Hasan.[80]

W8. Spinning and Weaving

The depictions of these activities are fairly abbreviated, with only three vignettes to illustrate the entire process.[81] These vignettes are arranged realistically, perhaps to represent the inside of a weaving shop. The fiber is hacked inside a shoe by one woman and spun by another; two additional women weave the spun thread into cloth on a horizontal loom.[82] The work in this atelier, which would have been located on the tomb owner's estate,[83] is supervised by "the Overseer of Weavers."[84]

---

75 *mḥ n kꜣ*

76 *sꜣw ꜥt rḥtyw Mꜥkt*, cf. Wreszinski, "Bäckerei," p. 5.

77 Cf. Drenkhahn, "Brot," LÄ I; and Helck, "Bier," LÄ I.

78 Cf. Barta, Opferformel.

79 See Bennett, "Growth of the *Ḥtp-di-nsw* formula."

80 See, for example, the west wall of Tomb 2 (Newberry, BH I, Pl. 12), and cf. Weeks, "Art, Word...," pp. 61-62.

81 The weaving scene here is discussed in great detail by Roth, in Ancient Looms, where he publishes all the available copies of this scene, concluding that those of Calliaud, Rosellini, and Davies are the most accurate (Roth, Ancient Looms, p. 7). See also Klebs, Reliefs II, pp. 127-130; Hagemann "Webestuhl," LÄ VI, cols. 1160ff.; and Johl, Webestühle, pp. 19ff.

82 Roth, Ancient Looms, p. 10; see also Garstang, Burial Customs, p. 132, for a model of a loom found in the lower cemetery at Beni Hasan.

83 Winlock, Models of Daily Life, pp. 29ff. and Pls. 25-27; Kemp, Anatomy, p. 154.

84 *imy-r mrt.*

Figure IV.9: Spinning and Weaving

The fiber used in spinning and weaving would normally have been flax,[85] a crop which is shown being harvested on the other side of the doorway on this wall;[86] the material produced would therefore have been linen. Such cloth, perhaps to be equated with or linked to the cloth being washed in microscene W2, could have served a number of uses in daily life, and would also have featured heavily within the context of the cult. For example, the burial would have been provided with various articles of clothing made from such material, and piles of uncut cloth might well have formed part of the funerary equipment; the mummy would have been wrapped in linen bandages. Any statue cult would also have needed such materials, since statues were ritually dressed each day.[87] This link is illustrated in Tomb 2, where a basket of cloth is carried by one of the attendants of Ameni's statue.[88]

W9. Building the Tomb?

There are eight vignettes in the lowest register to the south of the door. These depict various manufacturing activities which can be tentatively identified as connected with the building and dedicating of the tomb itself. Although there is no particular direction or sequence to these vignettes, a number of the figures face right, and the scene as a whole reads more coherently from right to left. The effect of the left half of the register is generally that of the inside of a crowded workshop, with the figures to the right forming a sort of procession.[89]

This "procession" is led by "the Scribe, Netjernakht,"[90] who holds a scroll in one hand and crosses his other arm on his chest in a gesture of respect. This

---

85 Strauß-Seeber, "Spindel, Spinnen," in LÄ V, col. 1156; Lucas and Harris, Materials and Industries, p. 142.

86 See below, microscene W11.

87 David, Religious Ritual, pp. 78-79.

88 See Newberry, BH I, Pl. 13.

89 I am grateful to Janet Richards for her suggestions about this scene.

90 *sš Nṯr-nḫt*. This man, or another man with the same name and title, appears again on this wall (microscene W11) and on the south wall at the cult meal.

type of gesture is generally directed toward someone; as suggested above,[91] the object of his homage may be the figure of Khnumhotep II who appears above the door, linked to him by the wide line which runs under this register, up the side of the door, and under Khnumhotep II. Netjernakht thus would appear to be reporting directly to the tomb owner.

Figure IV.10: Building the Tomb?

The substance of his presentation is suggested by the next several figures. First is a kneeling man who holds an object which is no longer recognizable due to poor preservation of this part of the wall; the way in which he holds his hands suggests that he may be measuring something. Behind this man is "the Architect,"[92] who holds a scribal palette and an ambiguous object (perhaps a plumb line?) in his hands. The next standing figure, "the Stone-Carver,"[93] holds out a rectangular grid. If this object is a cubit rod, an interpretation which is supported by the size of the calibrations,[94] it could have been used for measuring and laying out the dimensions of the tomb. These men could be interpreted as the architects or master craftsmen responsible for the tomb, and therefore either temporary or permanent members of Khnumhotep II's staff. The central figure could be the master architect, the "Stone-Carver" could be the master artisan responsible for cutting out the tomb, and the kneeling figure could be the master draughtsman responsible for the decoration of the tomb.

Between the architect and the stone-carver is a figure kneeling in front of a *nw* jar. A label above him reads *mꜣy*, which might be translated as a form of the verb "to see," however, the purpose served by this man is unclear. There are two other figures in this register who sit and lean over identical jars; these three figures could be interpreted as consecutive views of the same man or as three men in various stages of a single process. One of these men is provided with the title "Overseer of a Phyle, Netjernakht."[95]

---

91 See microscene W4.

92 *wḏꜥty*.

93 *ms-ꜥꜣt*, cf. Ward, Index, #803, although this example is not listed, and the spelling is different.

94 For actual examples of cubit rods, see Arnold, Building, pp. 251-52.

95 *imy-r sꜣ Nṯr-nḫt*.

The vignette in the center of the register shows a man polishing the chest of a standing statue. Most of the inscription above this figure has been destroyed, but what is left reads: "the moment when the god lives..."[96] Thus, although the process shown seems to be one of manufacture, the implication is that the statue is being imbued with life, as it would have been in an "Opening of the Mouth" ceremony. This statue is not identical to the statue shown being transported in microscene W1, but is close enough to it in general outline to be the same; alternatively, two different statues could be depicted. The two statues portrayed on this wall are also both identified as *nṯr*, "god." There is no way of deciding whether one or two statues are shown by these two images. The statue shown being manufactured here could have stood inside the tomb or in the context of a temple; here, if the overall interpretation of this scene is correct, it would seem that the tomb is being emphasized.

Another vignette in this register shows two men, one of whom is named Nakht, standing on either side of a small shrine with a rounded roof, apparently steadying it. Inside the shrine is the figure of a light-skinned man, probably an official or a priest, perhaps even Khnumhotep II's eldest son as chief priest. Above this man is the label *Ḥr mꜣꜥ ḫrw*, "Horus is justified," a label which appears several times in this tomb.[97] This phrase may refer to the kneeling figure as the justified Horus, or may be directed to the tomb owner, perhaps as represented by the adjacent statue. It can be noted also that the shrine here, like the statue, is not identical to the shrine in microscene W1.

The next vignette shows a man working over a frame, perhaps to be identified as a square level, in which case he would be associated with the tomb builders.[98] The last figure in this scene is "the Sealer, N..." who has pale skin and light-colored shoulder-length hair; he faces right, and holds a curved baton in one hand.

This microscene, taken as a whole, can at least be tentatively interpreted as representing the actual building and inspecting or consecrating of the tomb itself. The scribe who leads the group and the sealer who brings up the rear could be presenting the artisans and their work to the figure of Khnumhotep II above the door (microscene W4). The chief architect, the master stone-cutter, and the master draughtsman would therefore be presenting their work, i.e. the tomb, perhaps represented in miniature by the statue and the shrine, to their patron.

W10. Storage of Manufactured Goods and Grain

This scene is given an architectural setting. At the left is a doorway, and then a portico supported by three pillars with plant capitals. The area to the right shows a central pillared hall or porticoed courtyard, beyond which is a

---

96 *ꜣt ꜥnḫ nṯr....*

97 See above, microscene W1 and below, microscenes W17, S4, and Sh5.

98 For examples of this artifact, see Arnold, Building, p. 254.

granary building. The entire architectural complex is linked by a horizontal band which begins at the third pillar from the left and continues behind the second pillared hall.[99] This microscene can be divided into two sections: to the right is an unidentified building, and to the left is a granary complex.[100]

Figure IV.11: Treasury

Seated in the portico to the left is a figure in an enveloping cloak, which probably indicates that he is outside; he is labeled: "the Overseer of Sealers, Imat."[101] With Imat are two other men, one of whom weighs something on a standing balance while the other records the results. Such weighing scenes are usually seen in the context of metalworking,[102] and may serve here in part as a short-hand encapsulation of such a theme. It may also emphasize the careful control which Khnumhotep II's staff kept over his goods, perhaps here specifically relating to equipment for the burial cult. Such objects, for example the cloth, pots, and furniture which are shown being manufactured in other microscenes on this wall, might even have been stored in the building seen to the left of Imat.

In the other section of this microscene, several officials and scribes[103] supervise five men who carry basketfuls of grain from a pile in the central

---

99 This band, given the prevalance of perspective-like arrangements of items elsewhere on this wall, might indicate the base of the far wall of the court in which these figures are sitting. (Suggested by David Silverman.)

100 The storage of grain and manufactured items are shown together in Tomb 17 (see Newberry, BH II, Pl. 17).

101 *imy-r sḏ3wty 'Im3t.* This official also appears in the statue transport scene (see microscene W1).

102 See for example, the west wall of Tomb 2 (Newberry, BH I, Pl. 11); Drenkhahn, Handwerker, p. 29.

103 *imy-r pr n pr ḏt*, the Steward of the House of Eternity

sš Nṯr-nḫt, the Scribe, Netjernakht

imy-r pr ḏ3tt Nṯrw-ḥtp, the Steward of the estate Netjeruhetep. Several of these men may appear again elsewhere in the tomb: a "Steward of the House of Eternity" is seen on the south wall, "driving the cattle" at the cult meal; here he is named Niuꜥ. The "Scribe, Netjernakht" may be the same man who leads the procession in microscene W9.

room to the granaries.[104] Grain was the most important element of the Egyptian diet; it is shown being produced in several of the registers below (microscenes W11 and W12), and being processed on the other side of the door, in microscene W7.

Figure IV.12: Filling Granaries

It is very likely that the grain being stored here is destined specifically for Khnumhotep II's funerary estate. The primary evidence for this conclusion is the title of one of the officials seated in the pillared hall, "the Steward of the House of Eternity." The term "House of Eternity" *(pr ḏt)* has been interpreted in a number of ways. The word *per (pr)* can be translated either as "house," or as "estate;" in the context of a New Kingdom temple, it referred to the administrative estate belonging to the temple.[105] The term *per djet (pr ḏt)* is translated "dwelling place" (Wohnstatt) by Grdseloff,[106] while Drioton thinks it was a place near the settlement where the embalming of the dead was carried out, from which the funeral procession left for its journey to the tomb.[107] Wilson suggests that it was the "the property in this world which provided for existence into eternity."[108] Faulkner translates it simply as "estate."[109] Wilson's interpretation makes the most sense in the context of this tomb, which is itself designed to ensure the existence of the tomb owner for eternity. Certainly the word *djet* usually refers to enduring time, or eternity.[110] Thus it stands to reason that the "house/estate of eternity" would be linked in some way to a perpetual cult.

One important purpose of this scene was therefore to represent, and thus magically ensure, a carefully regulated supply of goods and grain, "set

---

104 As this scene shows, the granaries were filled from the top. This is seen clearly in a model from the Tomb of Meketre; it is also clear from this model that the windows here represent separate silos (see Winlock, Models of Daily Life, Pl. 20; model F; pp. 25-27).

105 Spencer, The Egyptian Temple, pp. 14-17.

106 Grdseloff, Das Ägyptische Reinigungszelt.

107 Drioton, Review of Grdseloff, pp. 1007-1008.

108 Wilson, "Funeral Services," p. 203, note 12.

109 Faulkner, Dictionary, p. 90.

110 Cf. Wente, "Funerary Beliefs," p. 22.

aside,"[111] as a label above one of the officials reads, for use in the funerary/divine cult. This scene also illustrates a very specific location within the Egyptian landscape: the courtyards and granaries found on Khnumhotep II's estate. Actual examples of granaries suggest that these structures would have been located within the walled compound of which the house was the central feature.[112] The environment here is therefore that of the floodplain. The season should be early *Shemu*, when grain was harvested.

W11. Harvesting and Threshing

There are four vignettes in this scene, from left to right: harvesting flax;[113] harvesting one type of grain; harvesting another type of grain; and threshing. The harvest was a very important season, as it was when crops were collected for use during the rest of the year. A good harvest was vitally important to the welfare of the country. The depiction of a successful harvest could serve magically to provide the tomb owner with a perpetual supply of grain for his funerary and daily cult meals; this grain would have been stored in granaries, as seen above (microscene W10), and made into various types of bread and beer (W7). The flax would have been used primarily in cloth such as is being produced in microscene W8 and washed in microscene W2.

Figure IV.13: Harvesting and Threshing

The environment indicated here is clearly the floodplain, in which fields of grain were located. A specific season of the year is also indicated here, namely early *Shemu* (March or April), when the grain was ripe and ready to harvest.

W12. Plowing and Sowing

There are three vignettes in this microscene. The first is plowing, done by two teams of cattle and men. Next is hoeing of the earth and sowing of seeds, carried out by three men. Finally, there are two officials, possibly connected with the re-apportioning of the fields after the flood, who stand

111 *ẖw ḥnw*

112 See Kemp, pp. 153-55. His evidence here comes primarily from Kahun and Amarna, with reference made also to the Meketre models.

113 *ẖtt mḥy*, "picking flax."

facing one another. These fields, which are harvested in the register above, would have been located in the floodplain. The time of year indicated is the late fall, the beginning of the season of *Peret*, when the inundation receded and the crops were planted.

Figure IV.14: Plowing, Sowing, and Officials

A label above the men wielding hoes reads "cultivating the swampy lake."[114] This may make reference to the fact that plowing and sowing were carried out after the recession of the flood, when the land was still somewhat swampy. The Creation, which is associated with the annual recession of the inundation, may also be implied here, although there is no specific textual link. One of the officials in the last vignette is called "the Steward of the Marshland, Netjernakht;"[115] the other official is "the General, Khnumhotep's son Nefer."[116] After the flood, it would have been important to re-allocate the re-emergent land properly, and these officials may have responsibilities connected with this task. This process of reapportionment is depicted in tombs of the New Kingdom,[117] and is attested textually in both the Old and Middle Kingdoms.[118]

W13. Gardening

The next microscene depicts work in various types of gardens, from left to right: a grape arbor, an orchard of fig trees, a vegetable garden, and a palm tree. Gardens were very important in the Egyptian economy. They played an important role in the daily nourishment of the Egyptian population: gardens furnished vegetables and fruits, and also some raw materials for the manufacture of everyday objects.[119]

The fruits of these gardens could all play a part in the funerary cult. The grapes would have been made into wine,[120] a beverage which is offered to Khnumhotep II at his cult meal and which also appears on his funerary

---

114 *skꜣ ḥn*. For *ḥnt* see Faulkner, Dictionary, p. 171.

115 *imy-r pr n sḫt Nṯr-nḫt.*

116 *imy-r mšꜥ H̱nm-ḥtp sꜣ Nfr.*

117 Klebs, Reliefs III, pp. 6ff.

118 Smither, "A Tax-Assessor's Journal," pp. 74ff.

119 Hugonot, Le jardin, pp. 3-5 and 225.

120 Wine-making is seen in several of the other Beni Hasan tombs, for example, Tomb 15 (see Newberry, BH II, Pl. 6)

menu.[121] Figs, the grapes themselves, and various vegetables (such as lettuce and onions, shown above the garden) from the vegetable garden also appear among the offerings heaped up on the tables on the south wall.

Figure IV.15: Grape Arbor, Fruit Trees and Vegetable Garden

An important element of a garden was a convenient water source; in fact, a basin of some sort is shown to the left of the vegetable garden. These gardens would probably have been situated on the grounds of Khnumhotep II's estate, perhaps close to the river, as would be suggested by the proximity of the river voyage below. The season indicated by the harvesting of the grapes is the late summer, in *Shemu*, when wine-making would normally have taken place.[122]

Hugonot sees the Egyptian garden as a microcosm, reproducing at a human scale the world of fertile nature within the bounds of civilization, and providing a contrast to the hostile world of the desert.[123] The regular sowing, growing, ripening and dying of vegetable life can be seen as a metaphor for the cyclical nature of all life.[124]

W14. Cattle Fording a Waterway

The theme of cattle fording the water is a favorite of artists in the Old Kingdom,[125] and appears in several of the other Beni Hasan tombs.[126] In this vignette, the heads of six cattle are visible; they are supervised by a number of herdsmen, some of whom are in a small skiff.

The best pasturelands for cattle were marshy areas,[127] which were found in the Delta and at the edges of the Nile Valley at certain times of the year.[128] It has been suggested that cattle herds from the middle and south of Egypt were taken to the Delta in the winter, after the recession of the flood, and then

---

[121] Cf. microscene S2.

[122] Lucas and Harris, Materials and Industries, p. 18.

[123] Hugonot, Le jardin, p. 169.

[124] Wildung, "Garten," LÄ II, col. 376.

[125] Klebs, Reliefs I, pp. 60-61.

[126] See Tomb 29 (Newberry, BH II, Pl. 28) and Tomb 2 (Newberry, BH I, Pl. 12).

[127] Ghoneim, Rindes, pp. 29-30; Helck, "Weidewirtschaft," LÄ VI, col. 1167.

[128] Butzer, Hydraulic Civilization, pp. 16-18.

herded back south during the summer or fall.[129] Thus the cattle might have been taken across small bodies of water such as are shown here on their way to and from seasonal pasturelands in the north of Egypt. Since the cattle shown as moving to the north, the season indicated would be the end of the inundation, the beginning of *Peret*. It is also possible that the cattle remained in the Beni Hasan area year-round, in which case the fording of waterways would have been part of the trip to and from the local pastures, possibly at the time of the cattle census, which is shown adjacent to this scene on the north wall.

Figure IV.16: Cattle Fording a Waterway

The waterway shown here may be a canal or tributary which could have cut across the floodplain. It could also represent a temporary body of water which might have collected at the low-lying fringes of the floodplain when the inundation receded; if this latter suggestion is correct, an identification of the season as late fall or early winter is again suggested.

Cattle were extremely important in the economy of ancient Egypt, and were cared for assiduously as valuable commodities. Domesticated cattle were used for agricultural tasks, and were kept for their milk and meat.[130] Cattle were slaughtered at the celebration of the cult meal (microscene S6), and were counted, as is the case on the north wall of this tomb, as taxes and as part of the funerary/cult estate (microscenes N6 and N10). With bread, beer, and fowl, cattle were one of the traditional elements of the invocation offering. An important aspect of this scene is the actual crossing of the water, seen as a dangerous activity, since the cattle would be threatened by aquatic creatures such as the crocodile which appears in this scene. On one level, therefore, this scene illustrates the proper care of these valuable animals, and guarantees their perpetual supply for the cult.

W15. Canal Activities

Several other activities are shown taking place in the same body of water which is forded by the cattle; these are primarily connected with fishing.

---

129 Ghoneim, Rindes, pp. 29-30; Helck, "Weidewirtschaft," LÄ VI, col. 1167.

130 Ghoneim, Rindes, pp. 101ff.

There are several vignettes here. In one, several men fish with lines;[131] in another, two men fish with a net while a young boy sits on the bank[132] and fillets a fish. There are two additional men in this register; their presence within this scene suggests that they are associated with the fishing, but their functions are unclear.

The best time for fishing in Upper Egypt was at the end of the flood, when the receding waters left pools in the fields[133] and at the edges of the floodplain. As was suggested for the cattle fording scene, the body of water represented here may be one of these pools. If this is in fact the case, the season of the year indicated is the late fall, when the inundation was receding. Again, as suggested for microscene W14, this body of water could also be a canal or tributary.

Figure IV.17: Canal Activities

Fish were an important element of the Egyptian diet, thus this scene may serve magically to ensure the provision of this food in the Afterlife. However, fish do not typically appear among the offerings presented at cult meals,[134] thus it cannot be assumed that this scene serves to contribute food to Khnumhotep II's funerary meal. The catching of fish may simply be symbolic of plentiful food in general, or may ensure supplies for servants or pets such as cats.[135]

W16. Riverine Voyages

The riverine voyages in this tomb, according to their captions, can be read as a sequence: a journey south to Abydos is to the right (north) of the entrance doorway, and the return trip from Abydos is on the left (south). As is common in Egyptian tombs, the two scenes, which are treated here as a single

---

[131] *sniḥ*, "the angler." Cf. Montet, Les scènes de la vie privée, p. 24.

[132] It should be noted that he is shown above the fishermen, indicating in a perspective-type view that he is on the far bank.

[133] Boessneck, Tierwelt, p. 118.

[134] See below, south wall.

[135] There are no cats shown in Tomb 3, but the owners of Tombs 15 and 17 had cats (see Newberry, BH II, Pls. 6 and 17), so it is likely that there were also cats in Khnumhotep II's household. However, I know of no specific evidence which would suggest that fish was used as food for cats other than the general taste of modern cats for fish.

microscene due to their close interdependence, are arranged so that the boats being sailed move south, with the prevailing wind, and the boats being rowed go north, against the wind but with the current.[136] There are four boats in these two scenes; three of these are the typical Middle Kingdom type, with high prow and stern, maneuvered with complex steering oars, and captained by a pilot who stands in the prow to sound the depth of the river;[137] the fourth boat is a cult barque.

The voyage to the right of the door shows one boat, under full sail, towing a second, and is labeled: "Sailing upstream in order to learn the requirements (*ḫrt*)[138] of Abydos, by the Hereditary Prince, Count, Overseer of the Eastern Desert, Neheri's son Khnumhotep."[139] The first boat is occupied mainly by sailors; the pilot stands in the prow and gestures. Also in this first boat is a cabin, hung with hide-covered sunshades, in which sit two figures wrapped in enveloping cloaks; these men are probably family members, perhaps sons of Khnumhotep II.

Figure IV.18: Voyage South

The second boat is in the shape of a papyrus skiff with upturned prow and stern, and can be identified as the cult barque. Such boats, containing coffins or statue shrines, are often seen in Old Kingdom funeral scenes being towed across the river to the cemetery.[140] The center of the boat is occupied by a lion-shaped bed, similar to the one being made in microscene W3, under a baldachin. On this bed is an anthropoid coffin. This coffin is shown with a

136 The two registers are not quite lined up across the doorway, with the register to the right falling slightly lower than the one of the left. This may be due to the wish to place the plowing and harvesting scenes, which occupy two registers to the right of the door, together, which required more space than was needed on the left.

137 Erman Life, p. 481-86; Vandier, Manuel V, p. 718.

138 Cf. WB III, pp. 318-19.

139 *ḫnt r rḫ ḫrt ȝbḏw in rpꜥ ḥȝty-ꜥ imy-r ḫȝswt iȝbty Nḥri sȝ Ḫnm-ḥtp.* Cf. Newberry, BH I, p. 68: "sailing up the river to perform the ceremonies at Abydos by the *ha*-prince Chnemhotep, son of Nehera."

140 Klebs, Reliefs I, pp. 41-42; Klebs, Reliefs II, pp. 57-58 and 137.

beard, and is identified by a small label as Khnumhotep himself.[141] The mummy is tended by a small boy and several officials, including "the Sealer, Netjernakht," a *setem*-priest, and a lector-priest named Nakhti.[142] By the head of the coffin is a small table laden with food; the thigh of an animal, bread, and beer are visible.

The *setem*-priest holds a censer over the mummy; next to him is the label: "Horus is justified."[143] As discussed above,[144] this label may refer to the priest, who may also represent the eldest son of the tomb owner, or may refer to the tomb owner himself, in this case, in the form of a mummy.

This scene appears to serve a number of purposes. According to the label, it represents a pilgrimage to Abydos, which may have taken place in reality during Khnumhotep II's lifetime and/or after his death, or may have been celebrated only ceremonially or symbolically. Abydos was an important center of Osirian worship, and there is evidence that pilgrimages were actually made to this site during the Middle Kingdom.[145] The primary purpose of such a journey seems to have been to identify pilgrims with Osiris, allowing them to share in the offerings made at the Osiris temple, and thus to renew their life-force.[146] It is possible that Egyptians were in reality taken to Abydos at a certain point after their deaths, either for the Great Festival of Osiris,[147] or perhaps for whichever festival corresponded most closely with their burial. If a journey to the Great Festival is implied by this scene, the season indicated would be the beginning of the inundation (*Akhet*).[148]

Another possible explanation for this scene is that it represents a symbolic trip to Abydos that may have been celebrated within the necropolis itself, with some part of the cemetery designated as Abydos. This might have been carried out as part of the funeral ceremonies, and would have served as a magical surrogate for the real pilgrimage.[149] The important event in any case is the identification of the deceased with Osiris (and possibly also with Horus), and the subsequent guarantee of rebirth.

The return from Abydos is depicted to the south of the doorway. In this part of the microscene, the two boats are labeled: "Coming and bringing the

---

141 *H̱nm-ḥtp.*

142 *sḏ3wty Nṯr-nḫt*
stm
ẖry-ḥbt Nḫti.

143 *Ḥr m3ʿ-ḫrw.* This label also appears in microscenes W1 and W9, where it is found in association with a lector-priest and an unnamed official, and again in the shrine. See above, microscene W1 for full discussion.

144 Cf. Heerma van Voss, "Sargtexte," p. 470.

145 Yoyotte, "Les Pèlerinages," p. 23.

146 Yoyotte, "Les Pèlerinages," p. 20.

147 For this festival, see Rundle Clark, Myth and Symbol, pp. 132ff.

148 Yoyotte, "Les Pèlerinages," p. 34.

149 Altenmüller, "Abydosfahrt," LÄ I, col. 47.

requirements of Abydos by the Count, Neheri's son, Khnumhotep, possessor of reverence."[150] Thus the purpose of the voyage to Abydos has been fulfilled. This appears to be illustrated graphically by a figure, dressed in an encompassing cloak, who is seated under a canopy in the prow of the boat; this figure can reasonably be identified as Khnumhotep II,[151] transformed from a mummy into a resurrected being of some sort.

Figure IV.19: Voyage North

Khnumhotep II is accompanied by a group of nobles, who are labeled as "the children of the Count," "the Overseer of the Sealers of Lower Egypt, Bakt," and "Buhapy."[152] Just in front of the steering oar are two more light-skinned men, with an associated label reading: "The Skipper, Khnumsankh."[153]

Standing in the prow of the second boat are three officials. In the center of the boat is a cabin, inside which the heads of five women, to be understood here as mourners,[154] in three compartments are visible. The labels associated with these figures indicate that Khnumhotep II is accompanied on this journey by his male and female children, and by his two wives, both of whom also appear on the north and south walls of this tomb chapel and in the shrine.

An inscription behind this second boat reads: "Words spoken: Welcome in peace!"[155] An expanded version of this spell appears in Tomb 2; this says: "Welcome in peace to your tomb of the necropolis with the praises of all the temples..."[156] This scene can therefore be identified at one level as part

---

150 *iit ḥr in ḫrt 3bḏw in ḥ3ty-ʿ Nḥri s3 Ḫnm-ḥtp nb im3ḫw*. Cf. Newberry, BH I, p. 68: "Coming to perform the ceremonies at Abydos by the *ha*-prince Chnemhotep, son of Nehera."

151 Cf. Altenmüller, "Abydosfahrt," p. 44.

152 *msw h3ty-ʿ*
imy-r3 sḏ3wt B3ḳt
Bwḥʿpy.

153 *nfw Ḫnm-sʿnḫ*

154 In later periods, the mourning wife was explicitly identified as Isis, strengthening the identification of the deceased with Osiris. (Roth, "Social Aspects," pp. 56-57.)

155 *ḏd mdw ii.wy m ḥtp.*

156 *ii.w(y) m ḥtp r is.k n ẖrt-nṯr m ḥst nt ḥwt-nṯr nbw...*

of the funeral procession; the reference to temples may imply participation in this event by the local temples at which these nobles were priests.

These two scenes, taken together, represent a pilgrimage to Abydos and thus illustrate the transformation of Khnumhotep II from a mummy into a resurrected being; they may also refer to an event connected with the tomb owner's funeral. The deceased is identified with Osiris, and perhaps also with Horus.[157] The major inscriptions associated with these two scenes state explicitly that the purpose of the journey is to "learn the requirements of Abydos." A number of religious and literary texts, examples of which date back into the Old Kingdom, provide evidence for a body of restricted religious knowledge which was originally reserved for royalty, but was gradually adopted by private people.[158] These "requirements" may be an example of such knowledge; they may represent some sort of specialized information that somehow enabled the deceased to be transformed.

Kessler also identifies the type of journey shown here with a primeval royal trip on the Sea of Pharaoh to the West, East, North and South, which he claims united the king as Horus with his father or mother gods and as Osiris with the royal ancestor gods (Re or Neith and Hathor). However, he provides no specific evidence for this claim.[159]

Another layer of meaning for these scenes is suggested by comparison with the voyages of the sun god in his day and night barques.[160] In the Egyptian conception, the sun god was born in the eastern desert, traveled across the water that borders the sky during the day, died each evening in the western desert, and traveled through the netherworld during the night to his rebirth each morning.[161] This interpretation is supported by the inscription found on an Abydene stela of the Middle Kingdom, dealing with a pilgrimage to Abydos, which reads in part: "...He rows in the night barque and sails in the day barque..."[162]

Several environments are therefore evoked in this microscene. At one level, the journeys to and from Abydos take place on the Nile. The orientation of these scenes within the tomb confirms this interpretation, since the boats literally travel south to Abydos and north to return. At another level, the funeral voyage, which could have been carried out either on the river or within the confines of the cemetery itself, may be portrayed. At yet a third level, the "river" here can be seen as symbolic of the waters of Nun surrounding the created Cosmos on which the solar barques were thought to sail.

---

157 Cf. Anthes, "The Origins of M3ʿ-Ḫrw."

158 Baines, "Restricted Knowledge."

159 Kessler, "Asiatenkarawane," pp. 156-157.

160 Cf. Altenmüller, "Abydosfahrt," LÄ I, col. 46.

161 Allen, Genesis in Egypt, pp. 5-6.

162 Yoyotte, "Les Pèlerinages," pp. 35-36.

OVERVIEW OF WEST WALL AS A WHOLE

This wall is extremely rich and complex, and its contents are many and varied. There are two primary themes underlying the decoration: preparation for the funeral and mortuary cult, and death and resurrection, as represented by the riverine voyages; the statue procession at the top of the wall in a way represents the culmination of these events. Taken as a whole, the wall magically ensures the transformation of the tomb owner into a resurrected being who will be maintained through cult activities. Possible identification of Khnumhotep II with both royalty and divinity is stressed here through the reiterated label "Horus is justified," and through the labeling of depictions of his statue(s) as *nṯr*, "god."

Many of the areas in the Egyptian environment are indicated here, from marshes and canals to the river to the floodplain, to the cult area itself, whether it be temple or tomb. If the wall is laid flat, the pattern which emerges reflects the geography of Egypt fairly accurately, as it shows the river in the center, flanked on either side by the floodplain, which contains both indoor and outdoor areas of the estate. The floodplain is bordered below by the marshes and canals, and also by the desert, implicit as the site of the tomb. In fact, if the actual orientation of the wall is maintained, the register containing the "tomb" (microscene W9) lies on the east bank of the river, as it does in reality.

Figure IV.20: Landscape Evoked on the West Wall

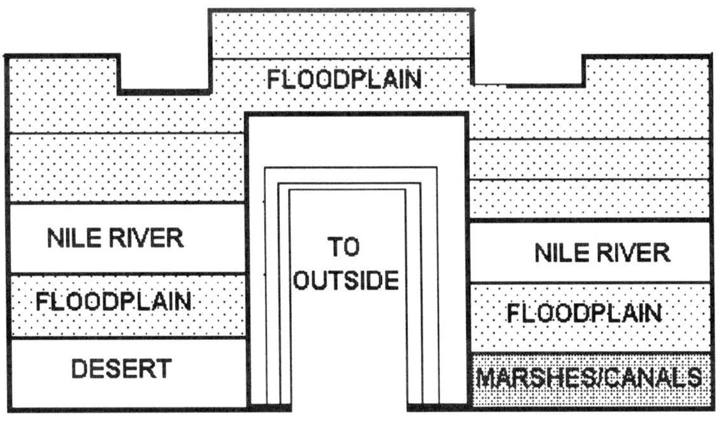

The seasonal cycle of the year is also implied on this wall, passing from the end of the inundation (end of *Akhet*), when the Great Festival of Osiris at Abydos was celebrated, to the late fall when the flood receded and the cattle were herded home, to the early winter (early *Peret*), when grain was planted, to the spring (early *Shemu)* when it was harvested, and finally to the late summer (late *Shemu),* when fruits were picked. The festival calendar of the Egyptian year is also implied by the statue transport scene. The cyclical nature of the Egyptian universe may also be stressed in the pilgrimage scenes, which can be interpreted as symbolizing the daily journey of the sun god through the sky and the netherworld.

Figure IV.21: Indications of Time on the West Wall

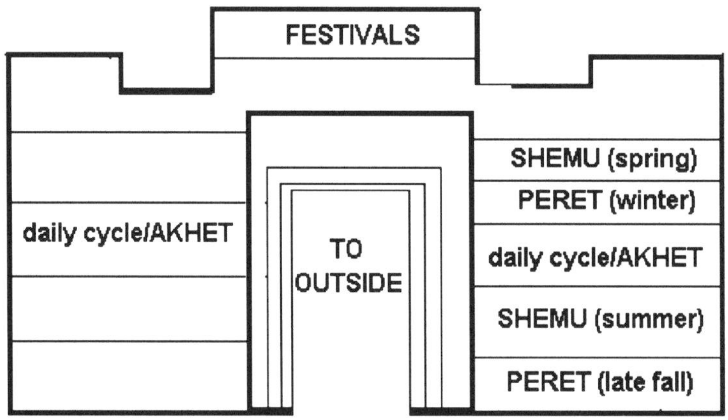

Many of the scenes on this wall fall into the category often termed "scenes of daily life." However, all of the scenes can be shown to be related to the funeral or the perpetual cult of Khnumhotep II, and should probably be thought of as providing instructions for a proper funeral and cult rather than illustrations of life on earth. The scenes, taken together, serve to ensure the proper transition of Khnumhotep II to divinity after death and provide both perishable and non-perishable goods for his Afterlife.

NORTH WALL

This wall is dominated by a large-scale standing figure of Khnumhotep II at the east end; most of the action here is directed toward this figure. The majority of the wall is devoted to the presentation to the tomb owner of various types of animals and birds; a census of these herds is recorded by a group of scribes. A second large-scale figure of Khnumhotep II is shown here hunting in a desert landscape. An illustration of this wall is given in Plate 2.

Figure IV.22: North Wall - Microscenes

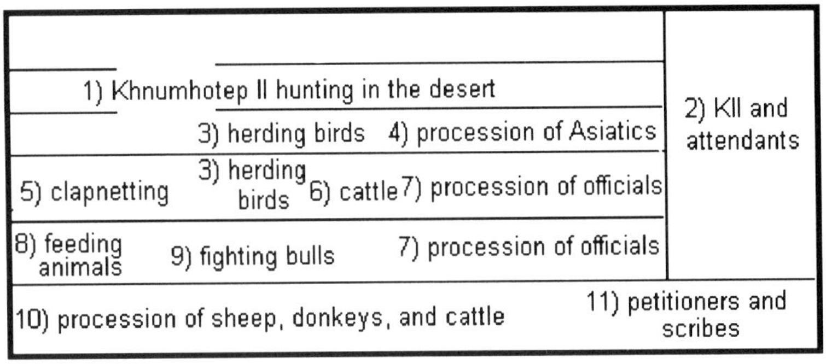

N1. Hunting in the Desert

The desert hunt focuses around a large-scale figure of the tomb owner (identified as: "the Count, Neheri's son, Khnumhotep, born of Bakt, justified"[163]) facing right and drawing a bow. He is accompanied by five of his sons. Four of them are in the registers in front of their father, each labeled with the title "son of the Count," followed by his name (they are Nakht, Khnumhotep, Neheri, and Netjernakht). A fifth son, "Khnumhotep, born of the (female) Sealer," is set off from the others on the top register, at the left (west) of the wall.[164] Khnumhotep II is also accompanied here by several dogs; a dog of the same species, perhaps even one of the same dogs, stands with the large-scale figure at the east end of the wall.[165]

A number of species of desert fauna are depicted here, including: a lion, hyenas, jackals, hedgehogs, mice, a hare, several wild cattle (that are the primary focus of Khnumhotep II's arrows), and a variety of other ungulates. Also seen in the desert landscape is a "griffin" with a feline body, long neck, and reptilian head; a human head flanked by wings emerges from this creature's back.[166] The environment portrayed in this scene is clearly the desert, which lies on the fringes of the Egyptian world. The season of winter (*Peret),* when young desert animals are born, is suggested in this scene by several animals who give birth.[167]

---

163 *ḥꜣty-ꜥ Nḥri sꜣ H̱nm-ḥtp ir.n Bꜣkt mꜣꜥ ẖrw.*

164 *H̱nm-ḥtp ir.n sḏꜣwtt.* He is probably separated from the others because his mother was secondary to Khety, the mother of the other four sons. See Ward, "The Case of Mrs. Tchat," for discussion of Khnumhotep II's putative second wife.

165 See microscene N2.

166 Similar animals appear earlier at Beni Hasan, in Tombs 15 and 17, in conjunction with other fantastic animals (see Newberry, BH II, Pls. 4 and 13 and te Velde, Seth, God of Confusion, pp. 15ff.). The human head is a new feature here.

167 Altenmüller, "Jagd," LÄ III, col. 222.

Figure IV.23: Desert Hunt

A large body of evidence, including arrow and spear points, Neolithic rock paintings, and desert animal bones recovered from settlement sites, demonstrate that the hunting of desert game was commonly practiced in ancient Egypt in the early periods of its prehistory and history.[168] This hunting probably served at first mainly to procure meat for food and animal by-products for various other purposes.[169] Many of the animals pictured in this scene could have served such practical uses. For example, the meat of the wild bull, like that of its domesticated cousin, could be eaten, and its skin could be used for clothing.[170] By-products of the Bubalis antelope were used in medicine, and its excrement was used as an aphrodisiac.[171] The skin of the

---

168 See Boessneck, Tierwelt, p. 36.

169 Altenmüller, "Jagd," LÄ III, col. 221.

170 Störk, "Rind," LÄ V, cols. 257-63.

171 Störk, "Antilope," LÄ I, col. 321; cf. Lucas and Harris, Materials and Industries, pp. 28ff.

dorcas gazelle was worked as leather, and its horns were used for handles.[172] The oryx, which is actually a type of antelope, was hunted for its meat, and its horns were used for bows.[173] The meat of the ibex was used for food, and by-products were used in medicine.[174]

Indications of the eventual use of these animals within the cult can be found in other microscenes within this chapel. Some of the animals captured or killed in the desert are enumerated on a papyrus presented to the second large-scale figure of Khnumhotep II on this wall by "the Scribe of the Offering Table, Renseb's son Mentuhotep."[175] Desert animals are subdued by several "children of the Count"[176] who are seen in a semi-register behind their father; such animals are force-fed several registers below (microscene N8). A number of the desert animals which appear here, including cattle, a gazelle, an oryx, an ibex, and some hedgehogs, are brought as offerings to Khnumhotep II on the south wall, and several desert animals are shown being sacrificed at the cult meal of Ameni in Tomb 2.[177] The heads of several desert animals also appear in the piles of offerings set before Khnumhotep II on the south wall. Clearly one important purpose of the desert hunt was to gather wild game for consumption at the cult meal.[178]

There is some question, however, as to the realistic nature of this hunting scene. Altenmüller suggests that by the Old Kingdom, when domesticated animals began to provide sufficient meat for the Egyptian diet, desert animals were hunted primarily in order to keep domesticated herds, cultivated fields, and human habitation safe from their depredations.[179] Over the course of the Old Kingdom, it is thought that the desert became more and more arid and inhospitable to life,[180] thus by the Middle Kingdom, it is possible that even the protective aspect of hunting would have become less pressing. A hypothesis offered by several scholars is that hunts such as are

---

172 Brunner-Traut, "Gazelle," LÄ II, cols. 426-27.

173 Störk, "Antilope," LÄ I, col. 321.

174 Störk, "Steinbock," LÄ V, col. 1276.

175 *sš wdḥw Rn-sb s3 Mntw-ḥtp*, see microscene N2.

176 *msw ḥ3ty-ʿ*.

177 See Newberry, BH I, Pl. 17.

178 Boessneck (Tierwelt, p. 36) notes, however, that the types of animals shown in tomb and temple reliefs do not correspond very well to the desert animal bones actually found in temple, cemetery, or settlement deposits. For example, the oryx, which is one of the most popular animals in the Beni Hasan tombs, is rarely found in archaeological contexts. He suggests, therefore, that the hunt shown is usually shown as idealized in tomb art, and that the animals generally depicted represent ideal offerings, rather than what was actually available.

179 Altenmüller, "Jagd," LÄ III, col. 221.

180 Cf. Butzer, "Wüste," LÄ VI, col. 1293.

portrayed here might have been performed primarily as sporting activities by this time.[181]

However, several features make it clear that this scene is not designed simply as a realistic portrayal of an event that Khnumhotep II enjoyed in life and wished to re-enact for eternity. The griffin, a completely imaginary animal, suggests that other levels of meaning are present in this scene, which is, in fact, full of symbolic references. Many of the animals who appear here have well-attested symbolic attributes which lend credence to other interpretations of the scene as a whole, and the sacrifice of these animals at the cult seems to have had meanings beyond simple nourishment.

The wild bull, which is the focus of the hunt for Khnumhotep II and his son and heir Khnumhotep, is highly symbolic of kingship and of attributes such as power and virility; at the same time, it can stand for the enemies of Egypt who must be subdued.[182] The lassoing of the wild bull is thought to have been an important ritual dating back to prehistoric times,[183] and the animal thus captured was a key sacrifice at the cult meal. The wounding of the wild bull is seen in the royal sphere as a complex statement: the king hunts a symbolically charged animal in which he himself can be embodied.[184]

The oryx, which appears on the standard of the Beni Hasan nome, was sacrificed by the Egyptian king as an offering to various gods. This rite appears in a number of Late Period contexts, but may go back to the Old Kingdom.[185] By Dynasty 25, the oryx was considered an incarnation of Seth;[186] it is also seen as the enemy of the eye of Horus.[187] Amulets in the form of the Bubalis antelope are not uncommon, and the head of this animal appears on the prows of various divine barques. The antelope head was thought to have apotropaic qualities, and ritual decapitations may date back to the Old Kingdom.[188] The dorcas gazelle was considered the animal of Isis in a cult at Coptos, and was the holy animal of Anuket. In mythology, the sun-eye turns into a gazelle at one point; this animal can also be identified with Seth.[189] The ibex was a holy animal of Min.[190] Several ungulates, specifically an ibex and a

---

181 Touny and Wenig, Sport, p. 74; Altenmüller, "Jagd," LÄ III, col. 221.

182 This attributes apply from at least the Old through the New Kingdoms, see Störk, "Rind," LÄ V, cols. 257-63.

183 Cf. Otto, "Hunting Ritual."

184 Altenmüller, "Jagd," LÄ III.

185 Derchain, Rites, pp. 10-14.

186 This identification may also date back to earlier periods; one argument against an earlier asociation, however, appears in the temple of Seti I at Abydos, where this nome is represented by an oryx although all mentions of Seth are supposedly eliminated (Derchain, Rites, pp. 10-14). See also te Velde, Seth, God of Confusion, p. 26.

187 Störk, "Antilope," LÄ I, col. 320.

188 Derchain, Rites, pp. 10-11.

189 Brunner-Traut, "Gazelle," LÄ II, cols. 426-27.

190 Störk, "Steinbock," LÄ V, col. 1276.

dorcas gazelle, are shown giving birth in this scene, thus overtones of fertility are added to the already rich symbolism of these animals.

The jackal was important as the embodiment of Anubis, as well as of Khentiamentiu, Wepwawet, Sed, the deceased king, and the souls of Hierakonpolis. This animal is seen primarily as protector of the dead, but also appears as a symbol of the king victorious over his enemies from at least the Early Dynastic period; the sun boat can be pulled by jackals.[191] The lion in this scene replaces the more commonly seen lioness.[192] The female of this species was very important in religion from the Predynastic period on, as an embodiment of several goddesses. She also represents the divine maternal agent at the daily rebirth of the sun. The male was a symbol of potency and power, and is seen as an embodiment of the king as early as Dynasty I.[193] The "griffin" is found in Egyptian art as early as the Predynastic period, and seems to have magical attributes. In later times, the griffin is called "the mightiest animal," and is thought to have had apotropaic functions; it is also a solar symbol.[194]

Although depictions of desert hunts are found in tomb decoration as early as the Predynastic Period and are common in the Old Kingdom, archers do not appear in private contexts until the late 6th Dynasty, and then only at provincial sites.[195] The royal version of this scene is attested first from the 5th Dynasty mortuary temple of Sahure.[196] However, the royal and private types of the Old and early Middle Kingdoms differ considerably in that the private tomb owner is never shown actively participating in the hunt in his own tomb, while the king is the primary hunter in royal contexts.[197] The appearance of the bow in private reliefs is seen by both Vandier and Altenmüller as part of the attempt of provincial nomarchs in the late Old Kingdom to accrue personal power by assuming royal prerogatives.[198] However, it is nameless hunters who use bows in these earlier scenes, as in Tombs 29, 15, 17, and 2 at Beni Hasan, not the tomb owner, as is the case in Tomb 3 and other contemporary tombs.[199] Thus, the "royal prerogative" of using a bow is apparently taken over

---

191 Barta, "Schakal," LÄ V, cols. 526-28.

192 Seen, for example, in Tombs 15, 17, and 2 (Newberry, BH II, Pls. 4 and 13; BH I, Pl. 13).

193 Rössler-Köhler, "Löwe," LÄ III, cols. 1080-90.

194 E. Eggebrecht, "Greif," LÄ II, col. 895; te Velde, Seth, God of Confusion, p. 21.

195 See Altenmüller, "Jagddarstellungen," LÄ III, col. 224.

196 There are also fragments of several desert scenes contained on the blocks reused at Lisht by Amenemhēt I; one of these may date from the 4th Dynasty, but would still post-date Rahotep and Nofret (the first private example). (Goedicke, Re-Used Blocks, pp. 134-136). For a summary and catalog of known examples of this type of scene in architectural contexts, see Vandier, Manuel IV, pp. 787-791.

197 See Borchardt, *sa3 hu-reʿ* II, Pl. 17, for a clear illustration of the royal type.

198 Vandier, Manuel IV, pp. 801-802; Altenmüller, Jagd, p. 23 and Altenmüller "Jagd-darstellungen," LÄ III, col. 230.

199 See Vandier, Manuel IV, pp. 743ff.

by people of lesser status before it is turned over to the nobility. In fact, it is very likely that the bow was in common use throughout the Old and Middle Kingdoms, and that the varying restrictions evident in the depictions of this scene reflect changing rules of decorum.

In this scene, therefore, Khnumhotep II has clearly assumed the role of the king. It is interesting that he does not wear the royal *shendyt (šnḏwt)* kilt, which is traditionally worn by the king in this context;[200] he does, however, wear a *shendyt*-like kilt on the east wall.[201] The fact that his sons imitate him takes this theme even a step beyond what is known from the royal realm.

When the desert hunt appears in a royal context,[202] it is interpreted as emphasizing the physical domination of the king over the chaotic power of the desert and its animal dwellers, and is compared to rites of the "destruction of the enemy."[203] The king as master of the hunt is identified with Horus, and the animals are symbolically identified with the enemies of the gods; the hunt is seen as the time and place where the ordered world of the Egyptian and the chaotic world of the desert come together.[204] Alliot has shown that, at least from the Ramesside period and later, royal representations of a desert hunt overtly symbolized the capture and destruction by the king of the enemies of Egypt.[205] It seems reasonable to hypothesize that private representations of the desert hunt, such as we have here, contain a similar message: the tomb owner, like the king, acts symbolically as Horus and is responsible for the maintenance of order in the universe. Khnumhotep II himself can therefore be said to be acting in a royal capacity, repelling *isfet* by subduing the chaotic forces of nature.[206]

The desert was also important in terms of solar symbolism: the eastern desert is the place where the sun rises, and the western desert is where it sets. As such, the western desert is the domain of Seth as the enemy of the sun god and the entrance to the netherworld.[207] The sun is also associated with the

---

200 Cf. Altenmüller, Jagd, p. 23. Other Middle Kingdom nobles do wear the *shendyt* in hunting scenes, e.g., Ukhhotep, nomarch at Meir in the early 12th Dynasty (see Blackman, Meir II, Pl. 8).

201 See below, microscenes E1 and E3.

202 The first example of a hunting scene appears in the Painted Tomb at Hierakonpolis (Quibell, Hierakonpolis, Pl. 78); hunting is also shown on several Predynastic ceremonial palettes (e.g., Emery, Archaic Egypt, Pl. 1b). The first royal attestation of this theme is found in the 5th Dynasty temple of Sahure (Borchardt, *Sa^{3} hu-reʿ*, Pl. 17).

203 Altenmüller, "Jagd," LÄ III, col. 232; Altenmüller, "Jagddarstellungen," LÄ III, col. 224.

204 Altenmüller, "Jagddarstellungen," LÄ III, col. 226.

205 Alliot, "Rites de la chasse."

206 Kessler also suggests such an identification for non-royal persons. He states that the deaths of desert animals are mentioned in royal liturgies and sees the eating of gazelle meat by Horus as an especially ritual act; however, he gives no references to specific liturgies or other potential sources of his statements. (Kessler, "Asiatenkarawane," p. 156.)

207 Butzer, "Wüste," LÄ VI, col. 1293.

injured eye of Horus, thus the desert and its inhabitants are the enemies of the eye and are again associated with Seth. The animals depicted reinforce this identification: the lioness presides at the daily rebirth of the sun, the jackal can pull the solar barque, and the gazelle can appear on the prow of this barque. The hedgehog and the mouse are both creatures of the sun god. Other animals are its enemies. Here the oryx plays this role, while in other of the tombs, the donkey and a fantastic animal identified with Seth can be present.[208] Thus one purpose of the desert hunt, at least as depicted in ritual contexts, could have been to ensure that the sun could rise every day and thus that the act of Creation could be perpetually renewed.

Since the entire scene is rife with solar symbolism, it can be read as a metaphor for the journey of the sun god through the netherworld. The dangerous animals which can be identified as enemies of the sun god suggest that the desert can be seen as symbolic of the netherworld or *Duat*, through which the sun must pass in order to be reborn each day. Khnumhotep II can here be identified as playing the role of the sun god, moving from west to east on his nightly path through the dangers of the desert; this identification also helps to ensure Khnumhotep II's own rebirth.

N2. Khnumhotep II and Attendants

Most of the north wall focuses on a standing figure of Khnumhotep II who faces left to receive the action that is directed toward him. This figure wears a long wig, a beard, a short kilt with a transparent overkilt, and sandals, and is carrying a staff and a handkerchief; he is accompanied by several dogs and his "Keeper of the Salve, Khnumhotep,"[209] who also appears with him on the east and south walls. The staff and sandals indicate that he is serving an official function and is outside.[210] He is also attended by "the son of the Count, the eldest son, whom God permitted to grow up Khnumhotep."[211] This man is actually his second eldest son, who is seen on the east wall and in the biography as his heir in the Oryx nome.

Above Khnumhotep II's head is the label: "Watching the levying of the cattle tax consisting of all animals (?, cattle, and donkeys), the tribute/dues *(inw)* which was brought to him from his towns and his districts of the interior of the Oryx nome and his city, by the Hereditary Prince, Count, Neheri's son, Khnumhotep, justified."[212]

---

208 See Tomb 17, north wall (Newberry, BH II, Pl. 13).

209 *iry-mrḥt H̱nm-ḥtp.*

210 Staehelin, Tracht, pp. 97-100.

211 *sꜣ ḥꜣty-ꜥ sꜣ wr sẖpr nṯr H̱nm-ḥtp.* (Cf. Fischer, Varia, p. 86.) This Khnumhotep is probably the second eldest son of Khnumhotep II, referred to in the autobiography as *ky wr*, "another eldest" (line 150; cf. Newberry, BH I, p. 43).

212 *mꜣꜣ irt irw m mnmnt nbt inw in n.f m niwwt.f spꜣwt.f nt ẖnw Mꜣ-ḥḏ niwt.f in rpꜥ ḥꜣty-ꜥ Nḥri sꜣ H̱nm-ḥtp mꜣꜥ ḫrw.* Cf. Newberry, BH I, p. 69: "Watching the produce, all the cattle and the

Figure IV.24: Khnumhotep II Receiving Census

'Inw is defined in the Wörterbuch as offering; tribute; gift; or product.[213] In an analysis of the use of this term in New Kingdom contexts, Bleiberg concludes that *inw* represented a system of gift-giving which was made to the king by both Egyptians and foreigners. Its purpose was primarily to maintain and emphasize the stature of the king as superior both to his subjects and to rulers of other lands; it also served to provide the king with a source of wealth separate from the redistributive economy of Egypt.[214] Another recent discussion by Spalinger of the occurrence of this word in a papyrus of the early 13th Dynasty suggests the translation "impost" for this term in certain

---

contributions brought to him from his villages and his fields of the interior of the Oryx nome, namely his estate (?), by the *erpa*-prince, the *ha*-prince, Nehera's son Chnemhotep, justified."

213 WB I, 91; see Bleiberg, "Privy Purse," p. 155.

214 Bleiberg, "Privy Purse," p. 167.

situations where it is clear that this is not a regular tax, but a response to special demands.[215]

Remembering that the context here is both pre-New Kingdom and non-royal, it is possible to see the application of both Bleiberg's and Spalinger's interpretations of this word to this scene. On the one hand, the materials being collected and counted can be regarded as taxes being collected for the king. This interpretation is supported by the presence of several royal scribes: "the Royal Document Scribe, Neferhotep,"[216] who presents the Asiatics to Khnumhotep II (see microscene N4); and "the Royal Document Scribe, Ameni,"[217] who helps with, and perhaps even supervises, the recording of the materials brought to Khnumhotep II (see microscene N11). Tax collection for the king at the cattle census is known to have been an important duty of the owner of Tomb 2 (Ameni), whose tomb contains a scene almost identical to this one.[218] This scene could thus function as eternal proof that Khnumhotep II was a good and loyal subject, an effective noble who carried out his proper duties and who helped to keep the country functioning properly.

It is also likely that this scene simultaneously represents the collection of *inw* for Khnumhotep's benefit, perhaps through the agency of the king, as emphasized by the presence of the royal scribes. This type of collection under royal auspices would be in keeping with the system of redistribution from the governmental center attested in various sources.[219] A comparable scene in Tomb 2 suggests that this collection is primarily for the benefit of the tomb owner's mortuary estate. In the earlier tomb, a scribe hands the tomb owner (Ameni) a papyrus bearing the following inscription: "Year...asses, totaling the writing of the livestock of the house of eternity, which come from the byre. Year 42...the Scribe, Khnumnakht's son Bakt, totalling ...30,000 asses, 3,000 cattle, counting stalled cattle of the house of eternity of the estate, by the Royal Scribe, Khnumnakht's son Bakt."[220] As discussed above,[221] the term *per djet* probably refers to the mortuary estate. The role of the king in setting up such an estate is attested from the Old Kingdom,[222] and the ultimate responsibility of the king for offerings is emphasized in the *hetep-di-nesu* formula: "An

---

215 Spalinger, "P. Bulaq," p. 209.

216 *sš ᶜn nswt Nfr-ḥtp.*

217 *sš ᶜn nswt 'Imny.*

218 This duty is described in Ameni's autobiography, where he claims: "...I was praised for it in the king's house in every year of the cattle tax *(irw)*. I delivered all their dues to the king's house..." (Lichtheim, Autobiographies, p. 139).

219 See Kemp, "Old Kingdom...," pp. 82ff., for discussion of this system.

220 On papyrus: *Rnpt ...ᶜ3 dmd sš n mnmnt nt pr d̲t pr m s3 / rnpt* 42 *t̲t n sš H̲nm-nḫt s3 B3ḳt dmd ᶜ3* 30,000 *iḥw* 3000 / *ḥsb md̲t nt pr d̲t md̲t nt ḥspt in...* (Plate 21)
sš nswt H̲nm-nḫt s3 B3ḳt (Plate 13). (Cf. Newberry, BH I, p. 32.)

221 See above, p. 104.

222 Kemp, "Old Kingdom...," p. 80.

offering which the king gives...to N." In a sense, Khnumhotep II himself also plays the role of the king here, by virtue of the fact that he is the recipient of this "tribute."

N3. Procession of Birds

In front of the hunting figure of Khnumhotep II, at the level of his feet, is a flock of twelve cranes, shown overlapping in a sort of melée. They are followed and supervised by a guardian, who holds a crane and his herdsman's crook in his arms. This theme continues in the register directly below, where a single herdsman has charge of three flocks of various birds, which can be identified as pigeons, ducks, and geese. Behind this man and his herds is another flock of fourteen waterfowl who mill around, some eating something on the ground, some raising their heads.

The birds shown here (cranes, turtledoves, geese, and ducks) are all excellent food birds.[223] All the birds shown can be domesticated, but are also caught in the wild in clapnets such as appear adjacent to the milling ducks, and kept as semi-domesticates.[224] Apparently wild cranes are not particularly good to eat, and must be tamed and fed grain before they are edible.[225] Thus these cranes could either be true domesticates, or tamed birds who are being fattened for the table like the birds in microscene N8. Both geese and turtledoves are also known as domesticates, and are excellent food; turtledoves are also tamed and kept as pets.[226]

Figure IV.25: Herding Birds

[223] Boessneck, Tierwelt, p. 97.

[224] Cf. Houlihan, Birds, #29, #32, #37, #42, #54.

[225] Houlihan, Birds, #42.

[226] Störk, "Gans," LÄ II, col. 374; Houlihan, Birds, #54.

These representations serve on one level to provide a perpetual source of food for the funerary cult. Poultry, especially ducks, geese, and cranes, are some of the most regularly seen items in the heaps of offerings at the funerary meal,[227] and appear on the funerary menu. By depicting the care and feeding of live birds, the continual supply of such meat for the deceased would be ensured.

N4. Procession of Asiatics

The right end of the third register contains a well-known depiction of a group of Asiatics, identifiable as such through their non-Egyptian hair styles and their brightly patterned clothing. This scene has been the subject of a great deal of discussion, and is probably the single most famous painting from this tomb.

Figure IV.26: Procession of Asiatics

The scene is captioned: "Arriving and bringing black eye-paint, which 37 Asiatics brought to him."[228] The group of foreigners depicted consists of four women, eight men, three children; they bring with them two donkeys, a gazelle, and an ibex. At the head of the procession are two Egyptian officials. The first of these men, "the Royal Document Scribe, Neferhotep,"[229] extends a sheet of papyrus to the large-scale figure of Khnumhotep II; on this papyrus is written: "Year 6 under the Majesty of Horus, Uniter of the Two Lands, King of Upper and Lower Egypt Khakheperre: the number of Asiatics which the son of the Count, Khnumhotep brought, on account of the black eye-paint; being Asiatics of Shu, number amounting to 37."[230] This inscription provides a historical date for this microscene, and perhaps for much of the wall. The second official is labeled: "the Overseer of the Hunters, Khety,"[231] a label

227 Vandier, Manuel, IV, pp. 12ff.

228 *iit ḥr int msḏmt in n.f ꜥꜣm* 37. Cf. Newberry, BH I, p. 69. The word for eye-paint is written with a quail chick *(w)*, which appears to be a mistake for the owl *(m)*.

229 *sš ꜥ n nswt Nfr-ḥtp.*

230 *rnpt 6 ẖr ḥm n Ḥr smꜣ tꜣwy nsw-bity Ḫꜥ-ḫpr-Rꜥ / rḫt n ꜥꜣmw in.n sꜣ ḥꜣty-ꜥ Ḫnm-ḥtp ḥr msḏmt m ꜥꜣm(w) n Šw rḫt ir* 37. Cf. Newberry, BH I, p. 69.

231 *imy-r nww Ḫty.*

which perhaps makes reference to the hunting scene above, or may refer to the desert animals brought by the foreigners.

Figure IV.27: The Chieftain, Ibesha

The foreigners are led by "the Ruler of the Desert, Ibesha;"[232] he and the man directly behind him present desert animals, (an ibex and a gazelle respectively), to Khnumhotep II. These animals tie into the desert hunt above and link these people to the desert. The men with Ibesha carry various un-Egyptian weapons, and one brings an asymmetrical lyre. This is the earliest evidence for this type of instrument in Egypt;[233] Hickmann suggests that it is part of a gift being offered to Khnumhotep II.[234]

This scene has been discussed by a number of scholars, whose views are summarized by Goedicke.[235] The general consensus has been that these Asiatics represent part of a caravan of traders who came to Beni Hasan to deliver the important commodity of eye-paint. It has been assumed that this event represented some sort of important trading coup of which Khnumhotep II, as its instigator or supervisor, was very proud. An alternative suggestion is that the scene shows Asiatics, accompanied by Egyptian border police, who have come to request permission to settle in the Nile Valley. More recently, these hypotheses have been challenged by both Goedicke and Kessler, each of whom reaches different conclusions about this scene.

Goedicke believes that thirty-seven people would be a very unlikely number for a trading caravan, and that the inclusion of women and children in such a group would be highly improbable. He also notes that Egyptians would not particularly need to trade for the commodity in question, since galena, the raw material from which the eye-paint was extracted, could be mined on Egyptian territory. As an alternative, he suggests that this was a group of working immigrants, moving to Egypt for the express purpose of working the galena mines. He interprets the two Egyptian officials

---

232 *ḥḳ3 ḫ3st Ibš3.*

233 Manniche, Instruments, p. 83.

234 Hickmann, Musikgeschichte, p. 28.

235 "Abi-Sha(i)," p. 203.

accompanying them as officials of the centralized government of Senwosret II; these officials were responsible for escorting the Asiatics to their temporary home in the mines. The expedition would have been therefore, in the final analysis, under the control of the government at Lisht. The reason for the placement of this scene in Khnumhotep II's tomb was, according to Goedicke, that Khnumhotep II as the ruler of Menat-Khufu would have had galena mines, presumably located in the desert to the east of Beni Hasan, under his jurisdication. Thus the scene would narrate not the delivery of black eye-paint, but the arrival of workers for the mines.[236] Goedicke also suggests that the "Ruler of the Desert, Ibesha" did not actually accompany these Asiatics to Egypt, but was pictured here because he was responsible for the activity in some way (perhaps as ruler of the territory from which they came). The man behind him would then be the actual leader of the group. He bases this belief on the assumption that the nomads would not literally have brought wild animals with them, so he instead reads these animals as elaborate rebuses.[237]

Kessler looks at this scene within the context both of the other scenes on this wall and of the tomb as a whole. In keeping with his generalized hypothesis as to the content of this wall, he sees the Asiatic caravan as part of the celebration of the *wep-renpet* festival. He agrees with Goedicke that a trading caravan is not represented, and that we are instead dealing with a group of migrant workers. He interprets the hieratic papyrus handed to the tomb owner as shorthand for a document which would have originally contained a register connected with the administration of the eastern desert, possibly for the purpose of listing supplies and payments owed to these workers.[238]

The question of who is meant by the "son of the Count, Khnumhotep" mentioned on the papyrus is also debated by these two scholars. Goedicke suggests that this "son of the Count," was an honorific title applied to Khnumhotep II at the royal court. Kessler, on the other hand, believes that "Count" refers to Khnumhotep II (as it frequently does in the other inscriptions in the tomb), and that "the son of the Count" is another son, Khnumhotep (who appears several times in this tomb and who stands behind his father in microscene N2). He hypothesizes that both Khnumhotep II and his eldest son Nakht have died by this time, and thus the caravan was presented to the second eldest son instead.[239] This assumption does not seem necessary, since we have no evidence that Khnumhotep II himself did not supervise the design and execution of his tomb paintings. There is also no reason why the younger Khnumhotep would put the representation of such an important event into his father's tomb rather than his own. It is more likely that the Khnumhotep

---

236 Goedicke, "Abi-Sha(i)," pp. 204-207.

237 Goedicke, "Abi-Sha(i)," pp. 207-209.

238 Kessler, "Asiatenkarawane," p. 149.

239 Kessler, "Asiatenkarawane," p. 149.

referred to on the papyrus is Khnumhotep II himself, whose father, Neheri, was also a count.

Kessler suggests that the scene represents an event necessary for the transfiguration of Khnumhotep II, performed at the *wep-renpet* festival, and designed somehow to link the tomb owner with Horus. He also suggests that this scene is related to the tribute of foreigners known from royal representations, for example, on the walls of the 5th Dynasty sun temple of Niuserre and in New Kingdom rock-cut Nubian temples, which he also interprets as depictions of the *wep-renpet*. Kessler states that in their royal contexts, at least in the New Kingdom, such foreign tribute scenes are part of the transfiguration of Horus, thus the inclusion of this scene puts Khnumhotep in the place of the king, and therefore Horus.[240] An identification of Khnumhotep II with both the king and Horus is attractive, given the other elements, but citations are omitted from Kessler's discussion, and thus renders it difficult to confirm.

Within the context of the other scenes on this wall, this procession could be seen at one level as simply representing another important item for Khnumhotep II's cult: eye-paint appears as one of the determinatives for the invocation offering formula in this tomb, and also is found on Khnumhotep II's cult menu. Thus the bringing of this item by the Asiatics ensures a continual supply of this valuable commodity.

There are other important levels of meaning that may be contained within this scene. Although the presence of a royal official suggests that the expedition illustrated here was carried out under the indirect supervision, and perhaps also partly for the benefit of, the king, who has delegated his authority to Khnumhotep II, the effect of the scene as a whole is once again to parallel royal iconography. Whereas a royal mortuary temple would have representations of bound prisoners to represent the king's domination of the earth outside of Egypt,[241] this chapel has the "caravan of Asiatics," which may carry a similar message of the subservience of foreigners.

The procession of Asiatic desert dwellers also ties in geographically to the desert hunt by which it is surrounded. The Asiatics are themselves desert dwellers. They are also foreigners, and therefore representatives of the semi-chaotic world which surrounds Egypt. This scene can therefore also be read as a statement of order over chaos: Khnumhotep II, whose two images on this wall physically bracket this tribe, has turned these non-Egyptian forces of disorder to his use, and to the use of the king and therefore of Egypt.

N5. Clapnetting

In the fourth register of this wall, the theme of the catching of wild animals contained in the desert hunt is echoed in a clapnetting scene. Two

---

240 Kessler, "Asiatenkarawane," p. 158-60 and notes 29-30.

241 For example, in the mortuary temple of Pepi II (see Jéquier, Pepi II).

clapnets are shown here. Each clapnet is pulled closed by two men; the nets are ultimately controlled, however, by a fifth figure, "the son of the Count,"[242] who holds the ends of both ropes. The two nets could also perhaps be thought of as a sequence: the net on top, in which the birds are more haphazardly arranged, is open; one man signals to the nomarch's son, and they all pull the net closed, resulting in the lower net, from which the captured ducks are removed. The man taking birds out of the lower net hands them to a second man, who ties their feet and hands them to a third, who hangs them from a yoke around his neck.[243] This scene feeds smoothly into the procession of birds being presented to the second large-scale figure of the tomb owner (microscene N3).

Clapnetting is a very efficient way of catching large numbers of birds at one time, and is an economical method of gathering waterfowl.[244] Many of the birds caught in these nets are now biannual visitors to the Nile Valley, stopping over on their migratory routes to the north and south; some are also breeding residents of Egypt.[245] It has been suggested that clapnetting was restricted to the papyrus swamps of the Delta and the Faiyum.[246] If this were the case, a majority of food birds consumed in the Upper Egyptian nomes would have had to be brought in from the Delta. Based on the modern ranges of the birds involved,[247] this assumption seems unnecessary.

Figure IV.28: Clapnetting

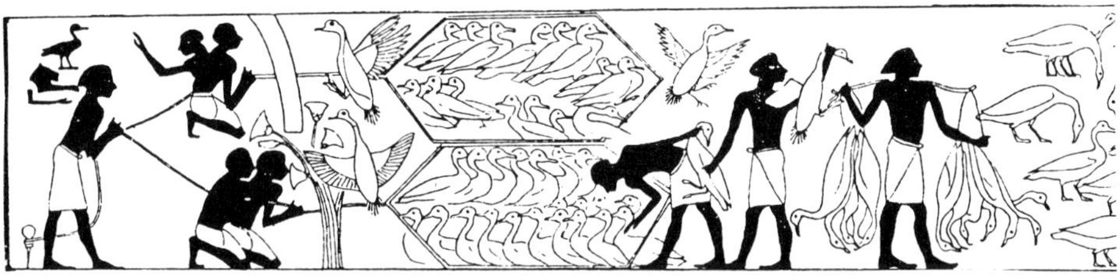

Since a number of the identifiable species in the clapnets are winter visitors to the Nile Valley, it is likely that this scene represents the season of *Peret*. Reference to this season is reinforced by the presence of a marsh; if this activity is indeed taking place in Middle Egypt, the period after the recession

---

242 *s3 ḥ3ty-ʿ*.

243 Vandier, Manuel V, pp. 363-64.

244 Boessneck, Tierwelt, p. 97.

245 See Houlihan, Birds. These birds can actually be identified from this painting, (see Houlihan, Birds, p. 69, figure 95). The birds in these nets include a variety of species; the contents of the upper net are more varied, while the lower seems to contain mostly pintail ducks.

246 Martin, "Vogelfang," LÄ VI, col. 1051.

247 For example, the pintail duck is a winter visitor to the Nile valley.

of the flood is the time when such marshes would appear. Some of the birds caught would have been killed immediately and either eaten or dried and salted for long-term storage, while others were tamed and herded for later consumption.[248] In this way, poultry for the year could have been collected.

Waterfowl were an important food source in ancient Egypt, and consistently appear as offerings,[249] on funerary menus,[250] and as determinatives in the invocation offering formula.[251] Their importance is also attested by archaeological remains of funerary meals[252] and finds at settlement sites.[253] A number of geese or ducks are brought to Khnumhotep II and Khety at the cult celebration on the south wall; of all the various foods included in the offerings at this celebration, both on the south wall and in the shrine, it is pintail ducks, examples of which are caught in these clapnets, that are presented to Khnumhotep II by his primary cult celebrant.

In addition to the procurement of food for the cult meal, this depiction of clapnetting expresses the concept of the maintenance of order over chaos. Birds could be seen, as for example in reliefs of the New Kingdom and later, as enemies of order:[254] the goose in particular seems to have been a symbol for the enemies of Egypt.[255] This aspect of the scene is underlined here by the fact that it is sons of the tomb owner who control the net, rather than nameless fowlers.

N6. Procession of Cattle

Behind the officials in the fourth register are several herdsmen with their cattle. Leading this group is one of the officials from microscene N7: "the Vice-Magistrate of the young of the short-horned cattle and the young of the long-horned cattle, Bakt."[256] He is followed by an unnamed herdsman who leads four cattle of various sizes on a leash. Bringing up the end of the group is "the Overseer of Sealers, Senankh."[257]

These cattle are above the animals being fed in microscene N8 and fighting bulls of microscene N9, and can perhaps be seen in some ways as the end result of these activities. They are perhaps representative of wild cattle

---

248 All of these activities can be depicted in association with clapnetting, for example in Old Kingdom tomb scenes; see Boessneck, Tierwelt, p. 99.

249 Vandier, Manuel, IV, pp. 93ff.

250 Barta, Opferformel.

251 Bennett, "Growth of the *Ḥtp-di-nsw* formula."

252 Cf. Emery, "Funerary Repast."

253 Boessneck, Tierwelt, pp. 94ff.

254 Cf. Kemp, Anatomy, p. 50, fig. 15 (Dynasty 19); Alliot, "Rites de la chasse."

255 Wildung, "Feindsymbolik," LÄ II, col. 147.

256 *nty m srt rnw wnḏw rnny iw3w B3ḳt.*

257 *imy-r sḏ3wty Sn-ʿnḫ.*

who have been tamed; in any event, they are destined to be counted as part of the cattle tax/census, and are important as sacrificial animals in the cult.

Figure IV.29: Herding Cattle

N7. Procession of Officials

The processions in registers 4 and 5 consist for the most part of male officials in calf-length kilts, each of whom salutes Khnumhotep II with his right fist to his chest. Each official is labeled with his name and title. The majority of the men are overseers of one sort or another, and may be connected with both the funerary and official estates of Khnumhotep II. This scene could serve, therefore, to ensure the proper functioning of these estates. The officials play much the same role in this scene as the herds of animals; they, too, are there to serve the tomb owner and ensure his eternal cult. Khnumhotep II also mentions in his biography that one purpose of his tomb was to "perpetuate the name of his official staff, they being established according to their offices;"[258] this function is clearly expressed by this scene.

Figure IV.30: Procession of Officials

[258] Lines 7-9: *srw.f rn n ḳnbwt.f smnḫw ḫft ỉ3t.sn.*

N8. Feeding Animals

The fifth register contains four vignettes which depict the feeding of various animals, namely birds, oryxes, goats, and cattle. The depictions are rather naturalistic, with the animals and their caretakers shown interwoven in realistic-looking ways. These vignettes have at least two possible levels of meaning. On one level, the feeding of the young[259] is an important part of animal husbandry, and could serve in the context of the tomb to ensure healthy animals for the perpetual cult.[260] On another level, they show the care which the humans take of their charges, and thus could symbolize the care which the tomb owner takes of those under his protection.

Figure IV.31: Feeding Animals

It is difficult to ascertain whether or not these various animals are true domesticates (that is, born in captivity) or tamed animals that have been caught in their natural environment and are being kept and fattened for the table and/or the cult. The birds are similar to those trapped in the clapnets in microscene N2, and oryxes are shown being caught and subdued in the context of the desert hunt. Wild cattle are also seen in the desert hunt, but domesticated cattle were common and very important in ancient Egypt. The goats were also probably true domesticates. These animals appear again on the south wall, where they are offered to Khnumhotep II and his wife Khety in the context of the cult meal.

N9. Fighting Bulls

To the right of the cattle-feeding are two vignettes which portray pairs of fighting bulls, each of which is exhorted and/or controlled by a herdsman. The actual purpose of such battles seems to have been to select the strongest bull for breeding purposes;[261] choosing the most aggressive bull to breed with the cows is sound animal husbandry, as it helps to keep the stock strong. These scenes are adjacent or very near to scenes of copulation in Tombs 29, 15,

---

[259] These animals are specifically labeled as the young (*rnn*) of their species.

[260] See Smith, "Animal Domestication," pp. 307ff.

[261] Ghoneim, Rindes, p. 36.

and 17, and close to cows suckling their calves in Tombs 15 and 2.[262] Thus an important purpose of this scene would be to ensure a perpetual supply of healthy cattle for the funerary estate and cult.

Figure IV.32: Fighting Bulls

N10. Procession of Sheep, Donkeys, and Cattle

Most of the 6th register depicts the bringing of various herds of domesticated animals to the census. There are herds of cattle, goats, donkeys, and sheep. Scribes stand in the midst of the herds of cattle and donkeys and write on tablets which they hold in front of them.

Figure IV.33: Procession of Herds

The animals in these herds are all important domesticates which were important in the Egyptian economy. All are potentially of use within the cult; cattle at least are brought live to Khnumhotep II on the south wall, and slaughtered in the context of the cult celebration. Some of the goats and donkeys are shown copulating, suggesting the season of spring *(Shemu)*, as well as a time of fertility and regeneration.

262 See Newberry, BH II, Pls. 31, 7, and BH I, Pl. 13. In Tomb 15 (BH II, Pl. 7) the nearby cow suckles a calf and a human child.

N11. Petitioners and Scribes

Directly below Khnumhotep II are several seated scribes who face left and record the herds which are brought to them. All of the scribes, one of whom is a "Royal Document Scribe,"[263] are labeled with their names and titles. Approaching these scribes directly is a group of three officials, one of whom, "the Keeper of Property, Netjernakht,"[264] holds another man with a staff around his neck. Netjernakht gestures with his left arm, while his prisoner bends over in supplication.

Figure IV.34: Petitioners and Scribes

The dispensation of justice often forms a part of census scenes; wrongdoers in such situations include herdsmen who have not presented their quota of animals,[265] which would seem to be the case here. By showing Khnumhotep II in charge of administering justice, this scene enhances his image as an effective noble embodying an aspect of royal responsibility. The prisoner, like the cattle, is brought in on a leash, as if he too were a subdued chaotic element.

OVERVIEW OF NORTH WALL AS A WHOLE

This wall, taken as a whole, represents the geographical environments of floodplain, marshes, and desert. The desert is evoked by the hunting scene and the procession of Asiatics; the floodplain, on which the estate of the noble was located, would have been the setting for the "tax" collection represented on the remainder of the wall,[266] and the clapnetting would have taken place in a marsh. These elements are arranged in imitation of the real landscape, with the marsh between the floodplain and the desert.

---

263 *sš ꜥn nswt.*

264 *iry ḫt Nṯr-nḫt.*

265 Montet, Les scènes de la vie privée, pp. 147-149.

266 See Winlock, Models of Daily Life, Pls. 13-16; model C, for a three-dimensional model of such a census. This shows a portico supported by vegetal columns under which Meketre (the tomb owner), his son and a group of scribes sit. Various herdsmen drive cattle past the portico while one man, presumably the chief herdsman, approaches Meketre.

The desert hunt evokes the season of the late fall/early winter (*Peret*), when animals give birth; the daily cycle of the sun, and perhaps especially the journey through the netherworld, is also indicated here. The most likely season for the census is the spring (*Shemu*), when the animals would be copulating. Support for this comes from the north wall in Tomb 2, which includes a depiction of the storage of grain in silos;[267] this grain would have been harvested and stored in the spring.[268] The date of Year 6 of Senwosret II found in the Asiatic procession provides a fixed historical point.

Figure IV.35: Landscape Evoked on the North Wall

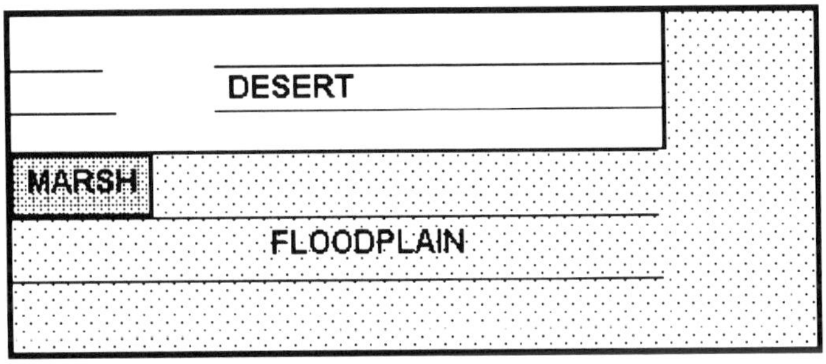

Figure IV.36: Indications of Time on North Wall

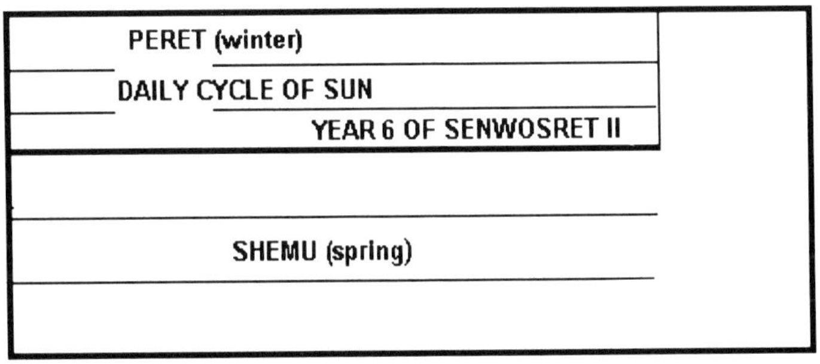

There are a number of complex statements contained within the decoration of this wall. Khnumhotep himself is both actor and recipient: he participates in the hunt and supervises the census of herds and officials and the arrival of a foreign delegation while simultaneously benefiting from these actions. One message of this wall is that Khnumhotep himself was responsible for his estate in life and therefore for the setting up of his funerary estate and cult before his death; that he can receive and enjoy this estate after his death is

267 See Newberry, BH I, Pl. 13.

268 See above, microscene W10.

his own doing. He also acts here as the supervisor of the local judicial system, represented by petitioners and scribes. In this context, Khnumhotep II is assuming royal functions delegated to him by the king's authority. He thus appears as an effective noble and a limited embodiment of the king.

There is a great deal of symbolism contained in specific features of the decoration of this wall. The hunting and clapnetting scenes, which involve the capture of wild creatures, represent the imposition of order by the tomb owner and his deputies (his sons) over the chaotic forces that intrude into the world of the Egyptians. These activities take place on the fringes of the civilized word: in the desert, and in the marshes. In the Asiatic tribute scene, Khnumhotep II's dominance over the chaotic area of the desert and the foreign lands is again stressed.

There are also allusions to fertility, virility, and regeneration within the decoration of this wall. The desert hunt contains animals giving birth and several particular animals associated with fertility. The hunting scene as a whole can be linked with the daily journey of the sun, which is itself representative of death and rebirth.

EAST WALL

The east wall is dominated by two large-scale figures of Khnumhotep II hunting in the marshes: to the north of the door he is fowling with a throwstick, to the south he is harpooning fish. Above the door, a slightly smaller-scale figure of the tomb owner is shown closing a clapnet. The lower registers of the wall contain scenes of seining (to the left) and fighting boatmen (to the right). An illustration of this wall is provided on Plate 3.

Figure IV.37: East Wall - Microscenes

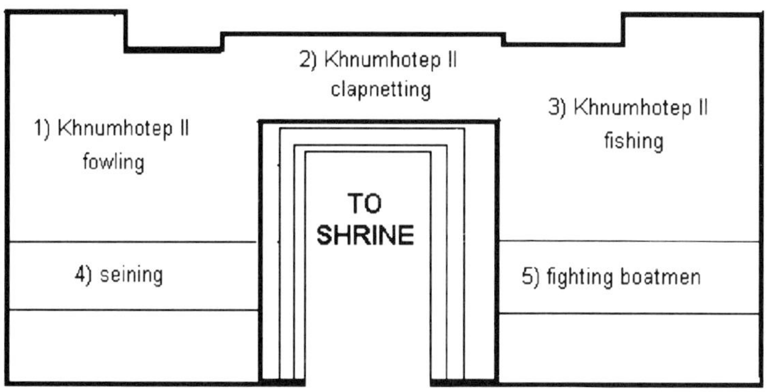

Around the doorway to the shrine are a number of inscriptions. On the lintel are four lines devoted to an expanded request for invocation offerings for the ka of Khnumhotep II; these lines include the most important of the tomb owner's titles and a list of festivals on which offerings were to be made. On the jambs are *hetep-di-nesu* prayers: to the left (north) of the doorway to Osiris, and to the right (south) to Anubis. Below each of these prayers is a seated figure of the tomb owner, dressed in a kilt, chest bands, and jewelry, holding a staff, and facing the axis of the tomb.

E1. Khnumhotep II Fowling with a Throwstick

To the left of the door is a large-scale figure of Khnumhotep fowling with a throwstick. Along the top of this part of the wall is the inscription: "the Hereditary Prince, the Count, Great One among the Nobles and Overlord of all the Nobles of the King, Neheri's son Khnumhotep."[269] Khnumhotep stands in a striding position, facing right, on a wooden platform inside a papyrus canoe whose prow is decorated with a *wadjet (wḏ3t)* eye; he wears a kilt similar to the royal *shendyt* kilt,[270] a pectoral, sandals, and short round wig. He also wears a long beard held on by visible straps, of a type which is usually associated with royalty and/or divinity.[271] The environment is a marsh, indicated by a band of water filled with fish, a hippopotamus, and a crocodile and a papyrus thicket filled with birds and their nests, butterflies, and a small genette. Khnumhotep II raises the throwstick behind his head with his right hand and holds three birds[272] in front of him in his left hand.

An inscription above Khnumhotep II's head reads: "traversing the swamps and marshlands by the Hereditary Prince, Count, Master of Secrets of the Sacred Writings, True King's Acquaintance, Overseer of the Eastern Desert, Intimate of the King, Foster Child whom he created, Foster Child of the King, whom he reared, the Count, Neheri's son Khnumhotep."[273] The emphasis on Khnumhotep II's relationship to the king here and in the frieze inscription is

---

269 *r-pꜥ ḥ3ty-ꜥ wr imy sꜥḥw ḥry-tp špsw-nswt nb Nḥri s3 H̱nm-ḥtp.*

270 The royal kilt has a long trapezoidal central piece, while the kilt seen here and in private representations of this hunt in the Old Kingdom has a triangular central piece (Staehelin, Tracht, p. 5 and Plate 20). For the royal kilt, see Jéquier, Frises d'objets, pp. 20-21 (compare to the non-royal kilts on p. 3). There is at least one example of a private person wearing the royal *shendyt* from the Old Kingdom (in three of the statues of S'ankhwyptah, see Staehelin, Tracht, p. 5), and a number from the Middle Kingdom (e.g., Ukhhotep, nomarch of Meir in the late Middle Kingdom, see Blackman, Meir VI, Pl. 13).

271 Cf. Staehelin, "Bart," LÄ I, cols. 627-28.

272 These birds can be interpreted either as decoys (Martin, "Vogelfang," LÄ VI, col. 1052), or as representatives of birds that have already been caught (Vandier, Manuel IV, cols. 720-21).

273 *ẖns š3w pḥww in r-pꜥ ḥ3ty-ꜥ ḥry sšt3 n mdw-nṯr rḫ nswt m3ꜥ imy-r ḫ3swt i3btt rḫ n nswt sḏty.f nswt sḏty šdt.f ḥ3ty-ꜥ Nḥri s3 H̱nm-ḥtp.* Cf. Newberry, BH I, p. 70; Montet, Les scènes de la vie privée, p. 10.

striking, especially given that he is wearing a kilt very similar to that worn by the king and a ceremonial beard.

Khnumhotep is accompanied by his wife Khety, who is identified as: "his wife whom he loves, priestess of Hathor, daughter of the Count, Lord [sic] of the house, Khety."[274] Also with Khnumhotep II is his Keeper of the Salve, Khnumhotep, who also accompanies the tomb owner in the harpooning scene across the doorway and on the north and south walls; he also acts here as Khnumhotep II's sandal bearer. Above the stern of the canoe is the "Sealer, Keeper of the property of her Lord, Tjat, born to Netjeru;"[275] this woman is thought to have been Khnumhotep II's second wife,[276] and certainly bore him a number of children, who appear elsewhere in the tomb.

Figure IV.38: Khnumhotep II Fowling

A fifth figure, identified as Khnumhotep II's son Khnumhotep, stands on his own ground line above the stern of the canoe, and is probably to be understood as standing on the bank nearby. He carries a large papyrus umbel

274 *ḥmt.f mrt.f ḥm-nṯr Ḥwt-ḥr s3t ḥ3ty-ʿ nb-pr Ḫty.*
275 *sḏ3wtt iry ḫt nb.s Ṯ3t irt.n Nṯrw.*
276 Ward, "The Case of Mrs. Tchat," p. 51.

in one hand and a handkerchief in the other. The inscription associated with him reads: "the son of the Count, of his body, to whom the heritage and rulership were given, Khnumhotep, justified, in Menat Khufu, when one appointed his son as ruler of Maat."[277]

It is probable that fowling with a throwstick was at one point a form of hunting practiced in Egypt.[278] This technique may originally have been a viable method for catching birds which was replaced by the more advanced and efficient technique of clapnetting. By the Old Kingdom, fowling with a throwstick is thought to have become a sport enjoyed by the royalty and nobility.[279] Textual evidence for this activity comes from a literary text whose composition is purported to be from the Middle Kingdom; this text describes the activity and lists the birds caught.[280] One reason often cited for the depictions of such scenes in tombs was to ensure magically that the tomb owner could enjoy in the afterlife a sporting activity that he had enjoyed during his lifetime.[281]

Activities of this type are usually assumed to have taken place in the Delta and Faiyum,[282] but it is equally possible that fowling in the Beni Hasan region could have been practiced in the low-lying marshy areas between the floodplain and the desert. These marshes would have been especially viable for such activities in the winter, just after the recession of the flood. Therefore, while one possible setting for this activity is the Delta or Faiyum, it also may have taken place nearer to home; alternatively, the function of this scene may be primarily ritual.

The ritual nature of the fowling is emphasized by the archaic nature of the weapon used.[283] That such ritual hunts were performed in the 18th Dynasty at Thebes by the king or his substitute is indicated by the fact that throwsticks used for the rite of fowling have been found in the sacred lakes of the Deir el-Bahri temple, and is also suggested by the fact that throwsticks have been found in royal tombs of Dynasties 18 and 19.[284] An important aspect of this action may be apotropaic: birds, identified as enemies of the creator god,

---

277 *s3 ḥ3ty-ʿ ẖt.f rdi.n.f iwʿt ḥḳ3t H̱nm-ḥtp m3ʿ ḫrw m Mnʿt-H̱wfw niwt.f rdi.tw s3.f r ḥḳ3 M3ʿt.* According to Khnumhotep II's biography, his eldest son, Nakht, was given a very high position in the Jackal nome, whence his mother, Khety came (Breasted, Ancient Records, p. 285), thus leaving Khnumhotep II's position for his second eldest (*ky wr*) son to inherit (line 150).

278 Martin, "Vogelfang," LÄ VI, col. 1051-52.

279 See Boessneck, Tierwelt, pp. 47ff. and p. 97.

280 Vandier, Manuel IV, pp. 746-47. The hunter is a Middle Kingdom king. See Caminos, Literary Fragments, pp. 22-39.

281 Touny and Wenig, Sport. p. 63.

282 Vandier, Manuel IV, p. 717.

283 Helck, "Papyrusraufen," LÄ IV, col. 672.

284 Säve-Söderbergh, Hippopotamus Hunting, p. 42.

are symbolically destroyed so as to maintain the proper order of the Egyptian cosmos.[285]

Fowling with a throwstick may also have represented the celebration of a cultic rite in the service of Hathor in her form as the goddess Sekhet, who has a special cultic homeland in the papyrus thicket.[286] Sekhet is actually mentioned in the inscription above a scene of Khnumhotep II harpooning fish, which is located on the other side of the doorway to the shrine. The connection between fowling and Hathor is emphasized in this fowling scene by the identification of Khnumhotep II's wife as a priestess of Hathor; this connection, and the presence of both Khety and Tjat, may also impart an erotic character to this scene.[287] This purported erotic character is enhanced by the activity itself, which involves the catching of waterfowl which themselves can have sexual connotations.[288] At least one New Kingdom fowling scene can be explicitly interpreted as containing erotic symbolism in the form of an Egyptian goose on the prow of the tomb owner's skiff.[289] Such an interpretation can perhaps be taken a step further: since love-making is intimately connected to procreation, the erotic overtones in these scenes can help provide magically for the continual rebirth of the tomb owner.[290]

This scene first appears in the 5th Dynasty funerary temple of Sahure, and is seen in private tombs dating from immediately afterward. It is generally assumed that the private scenes were based on the royal prototype, and that the depiction of this activity was originally a royal prerogative taken over later by private people. The royal origins of the scene are emphasized by the fact that the noble often wears the royal *shendyt* kilt,[291] or one similar to it, as is the case here. Kessler sees Khnumhotep II here as playing the role of Horus at the New Year's festival, ritually killing birds as well as hippopotami and crocodiles on the Sea of the Pharaoh as part of a Sacred Marriage with Hathor.[292] Although such a specific identification is difficult to prove, it does

---

285 See above, microscene N5, for references.

286 Martin, "Vogelfang," LÄ VI, col. 1052. Clear indication of the Hathoric content of these scenes is provided by a scene which can appear in connection with fowling (see for example, Tomb 15, Newberry, BH II, Pl. 4), where the noble shakes one of the papyrus stalks of the thicket in which the birds are nesting. When there are inscriptions associated with this scene, they label the action: *seshesh wadj (sšš w3ḏ)*, which is interpreted as "shaking the stalk" and linked with the sistrum. These inscriptions also mention Hathor. This scene is thought to have been connected originally to an action done in preparation for the actual hunt, wherein one shook the papyrus in order to scare the birds so that they would take flight. The shaking of the papyrus was originally offered to the "great holy mother," who is later identified with Hathor. (See Vandier, Manuel IV, pp. 738-46; Balcz, "Jagdfahrten," pp. 32-35.)

287 Derchain, "Snéferou," p. 22

288 Derchain, "La Perruque," pp. 62-64.

289 Derchain, "La Perruque," pp. 62-64.

290 Martin, "Vogelfang," LÄ VI, col. 1052.

291 Martin, "Vogelfang," LÄ VI, col. 1052 and note 13.

292 Kessler, Asiatenkarawane, pp. 155-156.

seem likely that Khnumhotep II is here functioning as an embodiment of the king or the creator god, acting to repel *isfet*.

E2. Khnumhotep II Clapnetting

In this scene, Khnumhotep II sits facing right on a feline-footed stool, wearing a knee-length kilt with a starched front panel and a lector-priest's sash.[293] He is behind a reed screen, and has just pulled closed a hexagonal net which is filled with birds. The inscription above him, which is just under the frieze, reads: "the Hereditary prince, Count, Overseer of the Eastern Desert, Neheri's son Khnumhotep, born of Bakt, possessor of reverence."[294]

Figure IV.39: Khnumhotep II Clapnetting

Behind Khnumhotep II are two officials. The first is "the son of the Count, his eldest, Nakht, born of Khety."[295] Nakht is also dressed as a lector-priest in this scene; he appears in the same costume again in the harpooning scene (E3) and on the south wall, where he is shown acting as his father's chief cult celebrant. The second man is labeled: "the Overseer of the Sealers, Bakt."[296]

The clapnet lies to the right of the reed screen, over a pool filled with waterfowl. Various water plants surround the net. The birds inside the net are carefully drawn, and can be identified as to species.[297] This scene is enriched

293 Cf. Staehelin, Tracht, p. 81.

294 *r-pꜥ ḥ3ty-ꜥ imy-r ḫ3swt i3btt Nḥri s3 H̱nm-ḥtp.*

295 *s3 ḥ3ty-ꜥ wr.f Nḫt ir.n H̱ty.*

296 *imy-r sḏ3wty B3ḳt.*

297 Cf. Houlihan, Birds, p. 68, figure 94; Boessneck, Tierwelt, p. 106.

by the addition of two acacia trees filled with birds, also identifiable as to species.[298]

This scene represents an important development in tomb iconography: the noble is demonstrating his power by doing single-handedly what it had heretofore taken at least two, and more usually four or more men to do.[299] Along with the seining and desert hunt scenes, clapnetting scenes in later temples could symbolize the capture of the enemy.[300] In these later scenes, the king or god closes the clapnet, as Khnumhotep II does here. He can therefore be compared to the king or a god, and interpreted as working to maintain order over chaos. The issue of collecting waterfowl for the cult seems to be of lesser importance here; perhaps this aspect is dealt with primarily by the clapnetting scene on the north wall.

E3. Khnumhotep II Harpooning Fish

The frieze text on this side of the wall is directly related to the scene below it: "the Hereditary Prince, the Count, Great one of the Catch, beloved of the Fen Goddess, Lady of the Catch, Khnumhotep."[301] The harpooning scene here serves as the companion to the scene where he uses the throwstick. The pose of the two figures is similar, as are the outfits (a transparent skirt has been added over the *shendyt*-like kilt[302] here) and the environments. In this scene, Khnumhotep II holds a harpoon[303] horizontally across his body in his outstretched arms, and has speared two *Tilapiae*. These fish are surrounded by an arc of water with a rounded top and vertical sides which is visible against the background of the thicket.[304]

---

298 All of these birds are beautifully drawn, and the characteristics of the relevant species are, for the most part, very accurately portrayed. In some ways, this scene could perhaps be considered a continuation of the tradition seen in Tomb 15 (and probably originally in Tomb 17), of long registers of carefully labeled birds (see Newberry, BH II, Pl. 4, and Davies, "Birds and Bats").

299 See, for example, Tomb 2 (Newberry, BH I, Pl. 12).

300 See above, microscene N5.

301 *r-pꜥ ḥꜣty-ꜥ wr (wrw) ḥb mry Sḫt nbt ḥb H̱nm-ḥtp.* Newberry, BH I, p. 70: translates this as "great in fish, rich in wild fowl, loving the goddess of the chase." Säve-Söderbergh, Hippopotamus Hunting, p. 38, has "great of fish and fowl, the beloved of the Fen-goddess, the Lady of the Chase."

302 The kilt worn by Khnumhotep II here is even closer to the royal prototype than the kilt he wears in microscene E1 (compare Newberry, BH I, Pl. 34 and Jéquier, Frises d'objets, p. 20, figures 50-52).

303 Bates ("Fishing," pp. 242ff.) defines a harpoon as having a detachable head with a rope attached, while the spearhead is fixed to the shaft; the harpoon is thrown, while the spear is thrust. It seems that the spear was used in the Old Kingdom, while the harpoon, which had been used in the Old Kingdom for hippopotamus hunts, was introduced into this scene in the Middle Kingdom.

304 This image, seen often in such scenes, has been frequently discussed (see Vandier, Manuel IV, p. 731-33); it is nicknamed the "mountain of water."

Above Khnumhotep is an inscription which reads: "traversing the swamps, bird pools, and marshlands by the Hereditary Prince, Count, Overseer of Fishermen of the Swamps and Bird Pools, Sportsman; I placed the screen; I caused the closing of the great net and the releasing of the harpoon; I speared thirty fish; how pleasing was the day of the shooting of the hippopotamus! The Count, Neheri's son, Khnumhotep, born of Bakt."[305]

Figure IV.40: Khnumhotep II Harpooning Fish

Khnumhotep II is again accompanied by his "Keeper of the Salve, Khnumhotep," who assists him by carrying an extra harpoon. Above the stern of the canoe is the "eldest son of the Count, Nakht," who also appears with his

---

305 *ḫns š3w sšw pḥww in r-pꜥ ḥ3ty-ꜥ imy-r wḥꜥw š3 sš wḥꜥ ḥb rdi.n.(i) k3[p] di.f sḫt n nmwt wrt wḥꜥ mrḏ rdi.n.(i) snt.s rmw.f* **30** *ḥs.wy stt db ḥ3ty-ꜥ Nḥri s3 Ḫnm-ḥtp ir.n B3ḳt*. Compare Montet, Les scènes de la vie privée, pp. 22 and 56-57: "Prenant des tas de gibier par le moyen d'une cachette," and "Il fait prendre au piege par le moyen du grand filet," "Il assure (?) le harpon...(?) le jour de harponner l'hippopotame;" and Newberry, BH I, p. 70: "Canoeing in the papyrus beds, the pools of wild-fowl, the marshes and the streams, by Chnemhotep, the chief canoer in the papyrus beds and pools of wild-fowl, capturing birds and fish: hidden by the screen he closes the great clap-net: spearing with the bident, he transfixes thirty fish; how delightful is the day of hunting the hippopotamus!" See also Save-Söderbergh, Hippopotamus Hunting, p. 21, esp. notes 3 and 4.

father in the clapnetting scene above the door. He once again wears a short kilt with a transparent overkilt and the band of a lector-priest. He holds three birds, perhaps as a reference to the fowling scenes nearby, or perhaps in his capacity of chief priest, since he also holds such birds at the cult celebration (see microscenes S4 and Sh5).

The reality of this type of fishing is attested, at least from earlier times, by harpoons with barbed heads of bone, horn, ivory, or flint found in Predynastic graves, especially in the Faiyum, and copper examples of such heads from 1st Dynasty royal tombs. It seems likely that spear- or harpoon-fishing originated as a practical method of catching fish. Actual examples of harpoon or spear heads are rare or non-existent from the Middle and New Kingdoms, and no examples of spear or harpoon heads designed to be hafted in pairs have been found.[306] Thus, at least by the Middle Kingdom, spear-fishing and harpooning do not seem to have been considered a particularly efficient way of catching large numbers of fish. The magical procuring of food for the funerary cult may have been one function of this scene, although this aspect was at best of secondary importance, since the catching of fish for food does not seem to have been a primary goal of the hunt,[307] and since fish are not generally thought to have been included at the cult meal.[308] Some scholars have suggested that these scenes were included in the tomb decoration in order to ensure magically that the tomb owner would be able to repeat a sporting activity in which he participated during his lifetime and which he wished to repeat for eternity.[309]

"Sport"-fishing in the marshes appears in two literary texts probably composed in the Middle Kingdom. In one of these a king, who has not visited the swamps in a while, goes into the Faiyum and sets out twice in his boat, once to spear fish and the other time to hunt birds. In the evening, he brings back an offering for the divinity of the area.[310] In this context, however, the reference might be ritual in addition to or rather than sporting. The other known literary mention of the fish hunt describes the fisherman as hiding in the shadows and catching numerous fish with each thrust of his spear.[311]

The earliest representation of fishing with a spear comes from a royal artifact of the Early Dynastic Period;[312] the next appearance is from the 5th Dynasty funerary temple of Sahure.[313] Private examples of this scene do not appear until the end of Dynasty 5, and thus post-date the royal examples.

---

306 Bates, "Fishing," pp. 233-40 and p. 243.

307 Säve-Söderbergh, "Harpunieren," LÄ II, col. 1014.

308 See microscene E4.

309 Touny and Wenig, Sport, pp. 64-66.

310 Caminos, Literary Fragments, pp. 22-39.

311 Caminos, Literary Fragments, pp. 1-21.

312 See Vandier Manuel I, p. 852, fig. 569.

313 Vandier, Manuel IV, fig. 399 I,2 on p. 721.

Depictions of harpooning fish are seen most often in conjunction with scenes of the tomb owner fowling with a throwstick,[314] as is the case here. Both of these activities are restricted to the tomb owner or his sons. It is generally thought, therefore, that representations of both spear/harpoon-fishing and fowling with a throwstick were icons restricted to the king in the early Old Kingdom, and taken over by high-ranking officials only later, in the 6th Dynasty.[315] This theory is given credence by the fact that the noble, when engaged in either of these activities, often wears a kilt reminiscent of the royal *shendyt* kilt, as Khnumhotep II does here. It has been suggested that the actual hunts were also exclusively royal, and that the king allowed his most faithful courtiers to participate in such hunts as a mark of favor.[316]

As was the case for fowling with a throwstick, harpooning fish may have taken place in the Delta or Faiyum.[317] Such a location for this scene is supported by the literary fragment cited above, and by an inscription from the sun temple of Neuserre which reads: "Coming to the Delta in order to catch fish in the cool water..."[318] The presence of the *Tilapiae* also supports this hypothesis, since these fish generally are found in the Delta. However, in other examples of this scene, the *Lates,* which prefers the environment of Upper Egypt, is seen alongside the *Tilapia,* thus it might be useful to explore interpretations of this scene on other levels. In fact, there are indications that this activity had become primarily ritualistic by the Middle Kingdom, and was linked with Horus and the king, and also with Hathor.

The type of fish, the *Tilapia,* which is shown as the quarry here would seem to be significant. This species of fish had important connotations of sexuality, rebirth and renewal, and served also as a protector of the dead. These connotations were apparently due to the odd breeding habits of this fish: the fertilized eggs are incubated within the mouth of the adult fish, so that the newborn *Tilapiae* emerge full-grown from the mouths of their parents. This fish also hides its offspring inside its mouth when there is danger present, thus the young can be seen to enter and leave their parents' mouths on a number of occasions.[319] Thus by depicting himself harpooning these fish, the tomb owner could represent the promise of regeneration, and allude in a number of appropriately decorous ways to the theme of sexuality. The sexual connotations of this fish appear more explicitly in a New Kingdom love poem, where a woman offers a*Tilapia* as a signal that she wishes to make love.[320]

---

314 Altenmüller , Jagd, p. 19.
315 Altenmüller , Jagd, pp. 16-18.
316 Vandier, Manuel IV, pp. 717-18.
317 Vandier, Manuel IV, p. 717.
318 Edel, Jahreszeitenreliefs I, p. 223.
319 Gamer-Wallert, Fische, pp. 60 and 129.
320 Derchain, "Symbols and Metaphors," p. 8.

The sexual connotations of this type of scene are emphasized in Tomb 29, where Tutqa accompanies her husband, Bakt I, into the marshes.[321] Tutqa is identified here as a priestess of Hathor, and holds a lotus, which is also a symbol of fertility and sexuality.[322] The link between the fishing and fowling in Tomb 3 also strengthens these connotations, since fowling, as is argued above, has clear Hathoric overtones. The papyrus shaking which helps to imply these themes for the fowling is also connected with the harpooning in a Pyramid Text (388a-b), which says in part: "I am the fish spearer, who comes from the sea, I am the papyrus shaker."[323]

There are also Hathoric references in the frieze inscription above the harpooning scene in this tomb, which identifies Khnumhotep II as "beloved of Sekhet, mistress of the catch." Sekhet, who was a personification of Hathor, is sometimes called "the Fen goddess" or "Lady of the Chase." She appears as early as the Pyramid Texts as a personification of the marshes, who provides the deceased with game to eat. In the context of the hippopotamus hunt, she appears as the divinity to whom the hunt is offered. The king can be described as "suckled by the Fen-goddess," an epithet presumably connected with the myths about the youth of Horus, which he spent hidden in the Delta marshes.[324]

Further support for the ritual overtones of the harpooning scene in Tomb 3 can be postulated from its resemblance to scenes of harpooning a hippopotamus that occur in royal contexts of the 1st Dynasty and Old Kingdom,[325] and then in private tombs of the New Kingdom and in temples of the Late Period.[326] These hippo hunting scenes have been carefully analyzed by Säve-Söderbergh, who concludes that the harpooning of the red hippo recreates the triumph of Horus over Seth, as is shown literally at the Ptolemaic temple of Edfu. The harpooning of the hippopotamus by the king symbolized the struggle with and victory over the powers of chaos.[327] This struggle also re-enacted the act of Creation, when *maat* prevailed over *isfet*, and was linked with rituals carried out before the enthronement of a new king.[328] This ritual action was connected with a feast called "Harpooning the Hippopotamus" that is mentioned on the Palermo Stone, and was celebrated as early as Dynasty 1;

---

321 See Newberry, BH II, Pl. 29.

322 Brunner-Traut, "Lotos," LÄ III, cols. 1091-96.

323 Helck, "Papyrusraufen," LÄ IV, col. 671. This quote also helps to link this scene with the Delta.

324 Säve-Söderbergh, Hippopotamus Hunting, pp. 37-40.

325 This is in contrast to representations of hippopotamus hunts in private tombs of the Old Kingdom, which are substantially different.

326 Säve-Söderbergh, Hippopotamus Hunting, pp. 14-21; see also Behrmann, Nilpferd.

327 Säve-Söderbergh, Hippopotamus Hunting, pp. 25-29.

328 Säve-Söderbergh, Hippopotamus Hunting, pp. 14-21 and 56.

this feast was evidently very important during the First Intermediate Period, at least at Mo'alla, and so was probably known at Beni Hasan.[329]

Säve-Söderbergh mentions Khnumhotep II's harpooning scene as an example of a scene based on the icon of the royal hippopotamus hunt, noting the hippopotamus in front of the skiff who looks back at the tomb owner and roars. He even suggests that the artist might originally have intended to depict a hippopotamus hunt and then changed his mind.[330] In fact, the label above this scene makes reference to "harpooning the hippopotamus."

This resemblance to the hippo hunting scenes strengthens the case for assuming important religious connotations for this scene. The case is also enhanced by comparison with its companion fowling scene, since there is evidence for temple rituals of fowling, and with the desert hunt scene, where the tomb owner again is the primary actor, and which can be interpreted as an icon for the triumph of order over chaos.[331] There is evidence from the Late Period that cultic bird and fish catching carried out by the king representing Horus served to guarantee the order of the universe.[332]

Thus the fish-spearing scenes can be said to have Hathoric connections, and, through the fish caught, help to protect the deceased and guarantee his rebirth. There may also be connotations of sexuality that emphasize the theme of regeneration, thus helping to guarantee the eternal rebirth of the tomb owner himself. At the same time, Khnumhotep II, in a royal guise, is shown symbolically maintaining order over chaos, and helping to reiterate the act of Creation.

E4. Seining

Below the scene of Khnumhotep II fowling with a throwstick is a second register, also bounded at the bottom by a band of water, in which ten men are seining with a large net. The men are divided into two teams, each of which pulls on a rope attached to one end of the net filled with fish. The seiners are supervised by "the Overseer of Fishermen, Mentuhotep,"[333] who faces the fishermen and gestures with his left arm. At the left end of this register is a single figure facing left. He has a yoke across his shoulders, with which he carries a large *Tilapia* and a basket filled with other fish. He is carrying these fish to the left, or north, with the result that he appears to be taking them to the census on the north wall.

---

329 Säve-Söderbergh, Hippopotamus Hunting, pp. 29-32.

330 Säve-Söderbergh, Hippopotamus Hunting, p. 8, note 21.

331 See above, microscene N1.

332 Gamer-Wallert, Fische, pp. 72-74 and135; Boessneck, Tierwelt, p. 47; see Alliot, "Rites de la Chasse."

333 *imy-r wḥꜥ Mnṯw-ḥtp.*

Figure IV.41: Seining

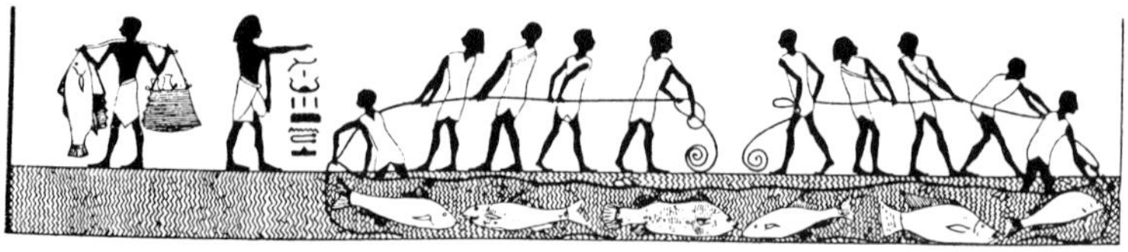

Seining is a very efficient way of catching large quantities of fish. The type of seine depicted here consists of a long strip of netting with parallel ropes at the top and bottom which meet in triangles at either side. Drag ropes are attached to the ends of the net; it is these ropes that the two teams of men pull.[334]

The fish inside the seining net are carefully drawn, and can be identified as species of fish that were still found in the Nile in recent times; all of these are from moderately to highly prized as food fish.[335] Fish were an important food source in ancient Egypt, thus one apparent purpose of these scenes would seem to be the ensurance of available food. There is some question, however, as to whether or not these fish are being caught specifically for consumption by the deceased tomb owner, since fish do not appear among Khnumhotep II's offerings or on his menus, nor are they presented directly to the tomb owner anywhere in this tomb. In fact, fish are never mentioned on funerary menus or seen in the piles of offerings at the cult meal in either Old or Middle Kingdom tombs and are only occasionally seen on the tables of guests, but never the tomb owner, in New Kingdom banqueting scenes.[336] It could be that fish were allowable as everyday fare but banned as food on festival days.[337] The exclusion of fish from cult meals could perhaps be seen as a consequence of their relative ease of attainment, which might make them less valued as food for a feast or for a noble in his Afterlife,[338] or a result of their sacred nature.[339]

---

334 Bates, "Fishing," p. 58-60. The nets were probably made out of twine spun from papyrus or other vegetable fibers, as is evidenced by the few actual remains that have been found (Bates, "Fishing," p. 62-63).

335 Boessneck, Tierwelt, Table 14, pp. 124ff. The species represented in the net include: *Gnathonemus cyprinoides; Mugil capito; Labeo niloticus; Barbus bynni; Mormyrus; Lates niloticus;* and *Tilapia*. See also Brewer and Friedman, Fish and Fishing, pp. 11-19.

336 Gamer-Wallert, Fische, pp. 66f.

337 Potential support for this idea is found in a calendar from the New Kingdom which lists specific days of the year, identified as festival days, on which certain fish are not to be eaten (Gamer-Wallert, Fische, pp. 75-84).

338 There are examples of cultures with taboos against fish that are based on their importance as a food source for the lower classes, thus their avoidance by the elite emphasizes the social stratification present. See Bates, "Fishing," p. 211. On the subject of taboos, see also Frandsen, "*Bwt* - Divine Kingship and Grammar."

339 See discussion below.

Alternatively, they may have served as eternal sustenance for the workers depicted in the tombs, or have been fed to sacrificial animals.[340] Thus, although at one level, the fish caught in the seining net can still be considered somehow representative of the abundance provided for the tomb owner in eternity, other explanations for the inclusion of this scene in the repertoire of the tomb decoration should be investigated.

One function of this scene is to represent the marshes and canals in which the fish would have been caught; this is in keeping with the other scenes on this wall. Also important is the time of year indicated. Assuming that a locale near Beni Hasan is implied in this scene, (which is more likely for this scene than for the more hieratic images above), if any specific season is meant, it is probably fall to winter, *(Peret)*, just after the flood. The recession of the flood waters provided an ideal situation for fishing in Middle and Upper Egypt, since at that time the fish would get caught in pools of water that were left behind in the fields.[341] This season is supported by the presence of the *Mugil* in this scene. This species of fish generally prefers the brackish waters of the Delta, but migrates south during the flood to breed, and could easily be caught after the flood, on its way back north.[342]

The presence of *Tilapiae* in the net, however, creates some difficulties, since in recent times these fish were found primarily in the Delta and Faiyum. One could theorize that *Tilapiae* might have traveled south during the flood, and been caught with the other fish in the stagnant pools left behind after its recession. This suggestion is, however, purely hypothetical; otherwise, it must be assumed that the fishing is taking place in the Delta or Faiyum, or that the primary reason for the presence of this particular type of fish is symbolic rather than realistic. The *Tilapia*, as discussed above (microscene E3), has strong connotations of sexuality and regeneration. These fish are also linked to the sun barque, as they were thought to accompany Re on his daily journey and protect him from his enemies.[343] The *Mugil*, with their yearly migratory breeding cycle, were also strongly linked to the cycle of Creation, and were considered manifestations of Re and Hapy.[344]

Fish in general can serve as protectors of the dead, a role clearly linked as early as the Middle Kingdom to the Osiris myth, in which various fish help Isis collect the pieces of her husband's body so that he can be properly mummified.[345] Through this Osirid connection, certain fish were thought to

---

340 See Gamer-Wallert, Fische, pp. 68-69.

341 Boessneck, Tierwelt, p. 118.

342 Boessneck, Tierwelt, pp. 121 and 130.

343 See Derchain, "Symbols and Metaphors," p. 8.

344 Gamer-Wallert, "Fische, religiös," LÄ II, col. 232.

345 Reference to this from the Middle Kingdom is found in an episode in the Ramesseum Dramatic Papyrus, where fish are sent out to look for Osiris (or the bits of his body), see Sethe, Dramatische Texte, pp. 118-19. This positive aspect of certain fish seems to have been

help protect and purify the deceased, and for this reason, mummies, especially of the Middle Kingdom, were adorned with fish amulets, and in the Late Period, fish mummies are common.[346] Thus another purpose of this scene might be to gather protection and guarantee regeneration and fertility for the tomb owner.

There is also evidence for negative connotations of fish. There is a body of evidence from the New Kingdom and later for explicit taboos against the eating of fish, especially by priests and other people for whom "purity" was an issue.[347] There are fewer such avoidances documented from earlier periods, aside from the absence of fish in the context of cult meals, but at least one spell from the Coffin Texts states that the deceased should not enjoy fish.[348] It has been suggested that some of these taboos could be attributed to an association of fish with the sea, which is undrinkable, and therefore seen as impure, or to the fact that they do not breathe air and were therefore seen as alien.[349]

Another potentially negative aspect of fish has been posited from the fact that they were left in pools in the fields when the flood waters receded each year, and thus may have been associated with the waters of chaos that preceded the Creation. Consequently, fish seem to have been identified with the enemies of the Egyptian cosmos. The earliest explicit example of this identification is in a magical text of the Ramesside period; other evidence comes, for example, from the Ptolemaic temple of Edfu, where there is a scene of fish being stomped upon, accompanied by the inscription: "Trample the fish, the enemy who is in the water; this is against evil."[350] Seining itself was seen, at least in the New Kingdom and later, as an image of the controlling of

---

reinterpreted in later times. Plutarch states that the Mormyrus, Barbus, and Mugil were believed to have violated the body of Osiris, yet there is clear evidence that these fish played positive roles both earlier and in the Late Period, and that their original function in the Osiris myth was to help collect the body of Osiris so that his resurrection could be accomplished. Thus Plutarch's statement could be a later explanation for the taboos surrounding these fish. (Gamer-Wallert, "Fische, religiös," LÄ II, col. 230-31.)

346 Gamer-Wallert, Fische, pp. 122-123 and136. See Mummies and Magic, p. 244, for examples of mummies.

347 For example, in a magical spell from the Ramesside Period which reads: "...You should not touch fish or goat, that which is forbidden to a god or goddess, because you are pure and your companions are pure." Another spell, from the Book of the Dead, says that it should only be spoken when both the deceased and his wife have refrained from eating desert animals and fish. The Triumphal Stele of Piye states clearly that fish-eaters were considered impure by the conquering Ethiopian Dynasty 25. Herodotus and Plutarch, along with other Classical sources, inform us that Egyptian priests were specifically forbidden to eat fish, and there are a number of other taboos, mostly local, attested from this period. (Gamer-Wallert, Fische, pp. 75-84; "Fische, religiös," LÄ II, col. 230. See also Brewer and Friedman, Fish and Fishing, pp. 12-13.)

348 Gamer-Wallert, "Fische, profan," LÄ II, col. 226.

349 Gamer-Wallert, Fische, p. 79.

350 Gamer-Wallert, Fische, pp. 72-74 and135.

chaotic forces, and the fish themselves can be identified as enemies of the sun god.[351]

This seining scene can therefore be interpreted on a number of levels. Certainly the magical gathering of food, even if not specifically for the cult meal, was probably an element of its purpose. Although the identification of fish with the enemies of the sun god is only explicit in later times, it is possible that such an identification was current earlier, and present here. If this is the case, then this seining scene could at one level symbolize the capture and subduing of the enemies of order. The positive aspects of some of the fish could help to provide protection and ensure the fertility of the tomb owner and his wife. The scene also evokes the watery aspects of the Egyptian landscape and indicates the end of the season of Inundation.

E5. Fighting Boatmen

The register below the scene of Khnumhotep II harpooning fish contains three papyrus skiffs filled with men who appear to be fighting with one another, trying to overbalance one another, or knock one another out of their respective skiffs. The locale is a canal or marsh filled with water plants, and the battle seems to be in earnest, as, for example, one man is suspended over the water between two of the skiffs. Another man is dancing, holding both hands above his head, and another, who stands in the water, is trying to get back into a skiff.

This scene is common in tombs of the Old and Middle Kingdoms, and is often associated with the gathering of papyrus. Touny and Wenig suggest that this was a sort of "shadow battle," put on as an exhibition contest, since in some examples the tomb owners are shown as spectators. Because it is clear in many cases that the boats are all being punted in the same general direction, it has also been suggested that this represents some sort of race, in which the various boat-loads try to beat one another home. There is evidence for similar games played by German fishermen in the 16th century and later.[352] Vandier interprets these scenes as play that took place after work in the marshes was finished for the day. In his view, the workers had energy left over after their day's work, and fought with each other to stretch their muscles and express their high spirits, and simply for pleasure. He interprets this particular scene as the depiction of the end of a typical conflict of this sort, with the adversaries reconciling. According to this interpretation, the dancing man is celebrating his victory, while the man being held between the skiffs is being saved from drowning.[353] In other examples, the fighting looks more serious, for example, when one of the combatants is being held under water.[354]

---

351 Doumas, "Fischer," LÄ II, col. 239.

352 Touny and Wenig, Sport, pp. 59-60.

353 Vandier, Manuel V, 512-13.

354 See Moussa, The Tomb of Nefer and Ka-hay, p. 26 and Pls. 10 and 11.

Figure IV.42: Fighting Boatmen

This scene and its predecessors at Beni Hasan appear to have two distinct phases: fighting, and the dancing. It seems possible, given the content of the rest of this wall, that there is some ritual meaning in this depiction. It could perhaps be connected to the well-attested battles between Horus and Seth.[355] In any case, the environment of a marsh or canal indicated is clearly in keeping with the other scenes on the wall.

OVERVIEW OF EAST WALL AS A WHOLE

All of the action on this wall clearly takes place in papyrus marshes or canals, as is indicated both pictorially and in the accompanying inscriptions. These marshes could have been located in the Delta, the Faiyum, or more likely in the Beni Hasan region, in the swampy areas between the desert and the floodplain.

Figure IV.43: Landscape Evoked on the East Wall

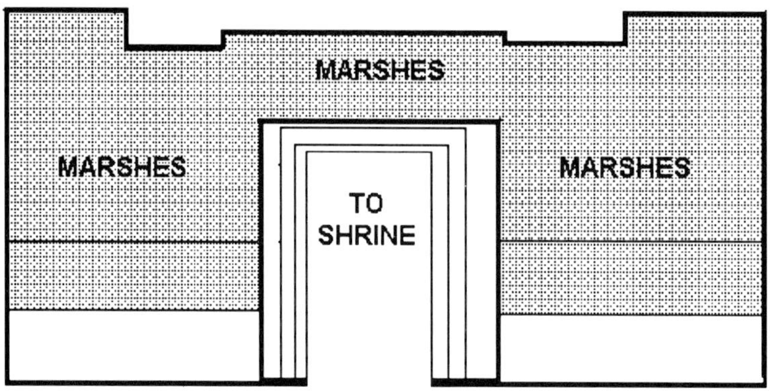

[355] Cf. Lichtheim, Literature II, pp. 214ff.

The fishing and fowling scenes suggest the season of winter, at the end of the inundation *(Akhet* to *Peret).* This period is evocative of the time of Creation, when the waters of Nun receded and the created cosmos emerged. The festival list above the doorway to the shrine should also be mentioned here.

Figure IV.44: Indications of Time on the East Wall

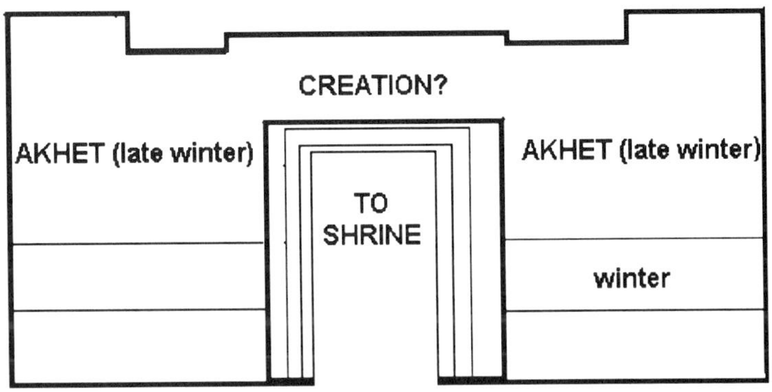

The images of fowling with a throwstick, clapnetting, and harpooning fish may not reflect real activities of Khnumhotep II. Although they can be interpreted as sporting events, and could well have been carried out either in the Beni Hasan region or in the course of visits to the Delta or Faiyum, their primary importance seems to be symbolic: they clearly show Khnumhotep II as the repeller of *isfet*, one who imposes order on the wild denizens of the marshes which lie at the fringes of the Egyptian world. These scenes show Khnumhotep II performing the roles of the king at one level and of the the sun or creator god on another. At the same time, they contain erotic references that reinforce the themes of rebirth and renewal. Taken as a whole, the wall reflects the theme of the perpetual renewal and proper maintenance of the created cosmos.

The placement of the harpooning and fowling scenes can be compared to the placement of similarly apotropaic depictions, for example the king smiting his enemies, on the pylons of New Kingdom temples. Such scenes usually show the king facing in, protecting the axis of the temple, as Khnumhotep II here guards the axis of the tomb and protects the shrine. These scenes also mark the limits of the profane world, stressing the sanctity of the shrine beyond. The addition of the clapnetting scene above the doorway to the shrine enhances this protective function.

SOUTH WALL

This wall is dedicated to the celebration of the cult meal of Khnumhotep II and his wife Khety. This cult celebration would have served to nourish and purify Khnumhotep II and Khety for eternity. The south wall is illustrated in Plate 4.

Figure IV.45: South Wall - Microscenes

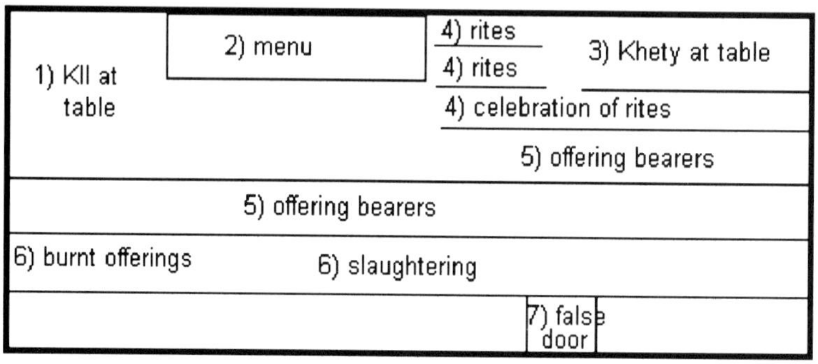

S1. Khnumhotep II at Offering Table

This microscene is located at the east end of the wall, and provides the main focus of the activities of the wall as a whole. Khnumhotep II is seated on a low-backed chair with feline feet on small pedestals. He wears a long wig and a short beard, a pectoral, bracelets, and a knee-length kilt with a starched panel in front. He holds a flail to his chest with his right hand, and extends his left hand toward the table in front of him. The presence of the flail is apparently a new feature in the Middle Kingdom.[356] This object is traditionally seen as a symbol of royalty or divinity, and is an especial attribute of Osiris,[357] thus its use here is significant, and supports an identification of Khnumhotep II with both the king and Osiris.

Above Khnumhotep II's head are a series of title strings, many of which are somewhat unusual, and some of which are only attested here:[358] "the Hereditary Prince, Count, Sealer of Lower Egypt, Sole Friend; the Hereditary

---

[356] Klebs, Reliefs II, 30 (note 24). The flail is seen for the first time at Beni Hasan in Tomb 2 (see Newberry, BH I, Pl. 7). This object has the typical angle of attachment that is characteristic of flails (cf. Fischer, "Geißel," LÄ II, col. 516 and Fischer, "Fächer," LÄ II, col. 83), and can be compared to flails found among the royal insigna on coffin friezes (see Jéquier, Frises d'objets, p. 187, figures 494-499, especially fig. 499, compare to flywhisks, p. 258, figures 680-684).

[357] Fischer, "Geißel," LÄ II, col. 516.

[358] See Ward, Titles.

Prince connected with the Sed; Count in the Upper Egyptian Shrine; the Hereditary Prince Connected to the White Chapel of Geb, Chief of Nekheb; the Hereditary Prince, Count, Lector-priest, Controller of Every Kilt; the Hereditary Prince, Count, God's Servant of Horus and Anubis, Neheri's son, Khnumhotep."[359] Behind Khnumhotep II is "the Keeper of the Salve, Khnumhotep,"[360] who holds an oversize unguent jar (the jar is at the scale of the tomb owner, while the official is at a smaller scale). This man also appears with Khnumhotep II on the north and east walls.

Figure IV.46: Khnumhotep II at Offering Table

The offering table consists of a round platter on a pedestal with a hollow foot. A low platform extends under the table and Khnumhotep II's chair; this may be a reed mat such as is seen clearly under Ameni's chair in Tomb 2.[361] An example of such a mat is carried by one of Khnumhotep II's attendants on

---

359 *r-pꜥ ḥ3ty-ꜥ sḏ3wty bity smr wꜥty*
r-pꜥ iry...wsḫt ḥ3ty-ꜥ m Pr-wr
r-pꜥ iry ḥḏ n Gb ḥry-tp Nḫb
r-pꜥ ḥ3ty-ꜥ sm ẖry-ḥbt ḫrp šndyt nbt
r-pꜥ ḥ3ty-ꜥ ḥm-nṯr Ḥr ꞽInpw Nḥri s3 H̱nm-ḥtp.

360 *iry mrḥt H̱nm-ḥtp.*

361 Newberry, BH I, Pl. 17.

the west wall of this tomb (microscene W4). On top of the table are fourteen large black-tipped reed leaves. Above these, arranged symmetrically, are waterfowl, onions, and loaves of bread. Below the table is written "the funerary meal"; also depicted here are a spouted jar and some fruits and vegetables.

To the right of the table are piles of food offerings, including poultry, cuts of meat and the heads of cattle and desert animals, fruits, vegetables, breads and cakes, beer, wine, and other foods and beverages. Many of the items illustrated here are shown being grown, collected, or prepared on the west wall, or caught and/or presented to Khnumhotep II on the north wall.

Figure IV.47: Piles of Offerings

S2. Menu

Above the piles of offerings in front of Khnumhotep II is a funerary menu. This contains fifty-four items, and is divided into two sections. The first part, which enumerates twenty-one items, lists the seven sacred oils, eye-paint, incense, natron, and a number of offerings and rituals. This section of the menu represents a somewhat abbreviated form of Barta's type B, which is basically an expression of the ritual of the Opening of the Mouth. This rite was performed on images of the deceased to enable them to receive offerings.[362] Entire meals, such as the "offering of the king" and "breakfast," are cited, and a number of phrases frequently seen in offering lists of the Middle Kingdom, such as: "sit down and be silent [at the invocation offering]!" are included here.[363]

The second section of the list is separated from the first by a line of inscription which reads: "Mighty One, who opens the chest, who brings the sacred oils."[364] There are 33 entries in this part, including libations, tools of

---

362 Barta, Opferformel, p. 78.

363 Cf. Barta, Opferformel.

364 *sḫm wn hn sḫp mrḥwt.*

various metals, the knife for the Opening of the Mouth, incense and natron; bread, water, grain, meat, onions, milk, and wine and beer.

Figure IV.48: Menu

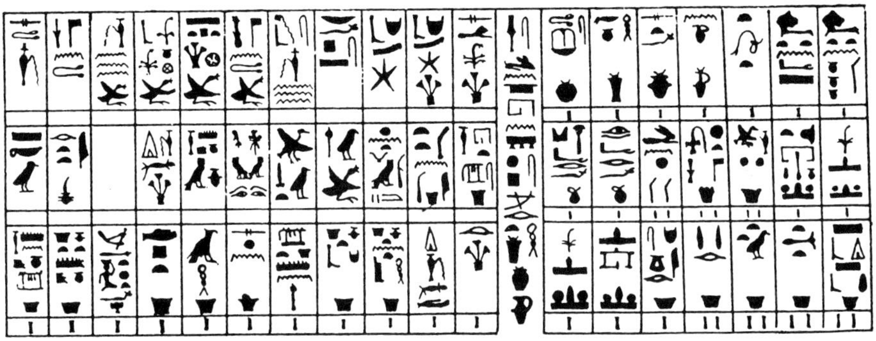

Since Khnumhotep II's wife Khety does not have a separate menu, this menu presumably is valid for both husband and wife.[365] The menu can be seen as a verbal expression of many of the foods and actions depicted on the remainder of the wall. It would have served as a back-up for the real rituals performed and offerings made to the deceased. The various substances listed, such as incense and natron, and the various rites described would have been thought to provide for the proper transition of the deceased to a being capable of receiving the offerings enumerated. Thus, by itself, this list could provide all that was needed for the transformation and perpetual sustenance of the tomb owner.

S3. Khety at Offering Table

At the west (right) end of the south wall, Khnumhotep II's wife, Khety, has her own separate offering table and pile of offerings. She faces left, and participates in her husband's meal rather than having a distinct cult of her own. Like her husband, she sits on a low-backed chair with feline feet which is in turn on a mat. She holds a lotus blossom to her nose; this flower is often seen as a symbol of regeneration.[366]

Khety's offering table is a round platter on a solid pedestal. It is ornamented with a checkerboard pattern and topped with thirteen reed-leaves, above which are two cooked geese. A reed mat piled with a variety of foodstuffs, including bread, meat, vegetables, and beverages, is located to the left of this table, slightly above the level of the table top. Above the mat and the table is inscribed: "invocation offerings to the ka of the revered one, the

[365] The lack of a separate menu is in contrast to the situation in Tomb 2, where both Ameni and his wife Hetepet have their own lists (see Newberry, BH I, Pls. 17 and 16).
[366] Brunner-Traut, "Lotos," LÄ III, cols. 1091-96.

daughter of a Count, Countess of the South, Priestess of Hathor, Priestess of Pakhet, Lady of the house, Khety, born of Tjenet."[367]

Figure IV.49: Khety at Offering Table

Behind Khety is a large unguent jar and a procession of five women, three of whom are her daughters (Bakt, Tjent, and Meres). The fourth is the (female) Sealer Tjat, daughter of Netjeru, thought to be Khnumhotep II's second wife,[368] and the fifth is a nurse named Rymat. Two small children flank Tjat; one is a boy named Khnumhotep and the other a girl called Satip. Ward suggests that these were the younger children of Khnumhotep II and the Sealer Tjat.[369]

Figure IV.50: Descendants of Khnumhotep II

---

367 *prt-ḫrw n k3 n im3ḫt s3t ḥ3ty-ʿ ḥ3tt-ʿ rsyt ḥm-nṯr Ḥwt-ḥr ḥm-nṯr P3ḫt nbt pr Ḫty irt.n Ṯnt.*

368 Ward, "The Case of Mrs. Tchat."

369 Tomb 4 at Beni Hasan consists only of a portico dug out of the rock but, according to all indications, was planned as a large tomb of the same type as Tombs 2 and 3. It bears an inscription on its lintel assigning it to "the Hereditary Prince, Count, Khnumhotep, born to the Lady of the House Tja(t)" (see Newberry, BH I, p. 7). Ward suggests that this Khnumhotep was the son of Tjat and Khnumhotep II, and that Tjat and Khnumhotep II were married after the death of Khnumhotep II's first wife, Khety. (Ward, "The Case of Mrs. Tchat, pp. 56-57.)

S4. Celebration of Cultic Rites

A number of cultic rites are depicted in the three top registers of this wall. The principal celebrant is Khnumhotep II's eldest son, Nakht, who is identified by an inscription which reads: "the eldest son of the Count of his body, who caused that inheritance be given to him, when the sun rises, Nakht, justified, possessor of reverence, born of the Lady of the house, Khety."[370] Nakht faces his father, wearing the short kilt, transparent overkilt, and lector bands in which he was seen on the east wall;[371] here he has also added an elaborate belt. Before him is a pile of offerings and two stands for burnt offerings, and facing him is an offering bearer who carries a basket of cloth or grain. Nakht holds several pintail ducks in his hands. The eldest son, as the heir of his father, was traditionally the chief celebrant of the mortuary cult, as the king, as the heir of the gods, was the chief celebrant of a divine cult. The son would thus ensure his succession to his father's position[372] and the continuance of his father's cult.

Nakht is followed by three more sons of Khnumhotep II and Khety: Khnumhotep, Neheri, and Netjernakht. A second Neheri, the son of Khnumhotep II and Tjat, brings up the rear of this procession.[373] The four sons of Khnumhotep II and Khety also appear on the north wall, hunting, and Khnumhotep and Nakht also hunt with their father on the east wall. Tjat's son Khnumhotep also hunts on the north wall, separate from his half-brothers.

A number of rites are depicted in the registers above the figure of Nakht, and continue behind the procession of the lesser sons. These include all of the important elements of the funeral/cult celebration: the pouring of a libation by a ka priest, the reciting of spells by a *setem*-priest and several lector-priests, the rite of "removing the footprint,"[374] the offering of incense, and the presentation of the sacrificial thighs. The rites in the first two registers illustrate the key elements of the offering ritual, and mirror key entries in the menu (microscene S2), beginning with the purification and ending with the removal of the footprint.[375] The actions of the priests and officials in the

---

370 *s3 ḥ3ty-ʿ smsw n ẖt.f di di(w) n.f iwʿ rʿ ḫʿ Nḫt m3ʿ ḫrw nb im3ḫw ir.n nbt pr Ḫty.*

371 Microscenes E2 nd E3.

372 However, Nakht in fact succeeded to the position of his maternal grandfather in the Jackal nome, while his brother Khnumhotep was to assume his father's duties at Beni Hasan.

373 Tomb 360 in the lower cemetery at Beni Hasan belongs to a "Neheri, born to Tjat," who has been identified by Ward with this figure. Ward suggests that Neheri died before the putative marriage of Khnumhotep II and his mother, accounting for his "mediocre" burial and lack of titles. (Ward, "The Case of Mrs. Tchat," pp. 52ff.) The fact that he does not appear on the north wall might suggest that he died before this wall was decorated; this hypothesis, coupled with the fact that Tjat's son Khnumhotep is seen as a child on this wall and an adult on the north wall, indicates that the south wall may have been painted before the north wall.

374 This rite has been reinterpreted by Altenmüller on the basis of a retranslation of the label as "bringing of the sacred water," see "Eine neue Deutung," pp. 152-53.

375 See Lapp, Opferformel. pp. 153ff.

procession behind Khnumhotep II's sons also make reference to menu entries, and may be connected with the Opening of the Mouth Ritual.[376] One of the lector-priests is labeled with the phrase that appears three times on the west wall: "Horus is justified!"[377] This phrase suggests that either Khnumhotep II or his son Nakht is again identified here with the god Horus.[378]

Figure IV.51: Nakht and the Performance of Rituals

Also included in this procession is a woman named Redyttepet and her two daughters; the placement of these figures suggests that they are relatives of the tomb owner. Redyttepet might even may be identifiable as yet another "wife" or mistress of Khnumhotep II, or she could be a sister.

Figure IV.52: Other Sons, Priests and Officials (and Third Wife?)

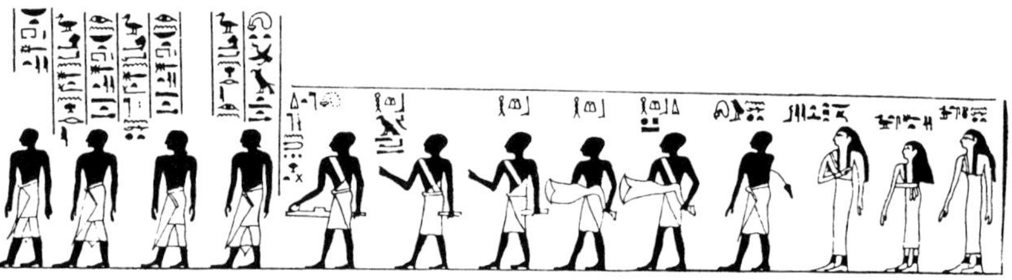

[376] Grieshammer, "Mundöffnung(sritual)," LÄ IV, cols. 223-24.

[377] *Ḥr mꜣꜥ ḫrw*. See also microscenes W1, W9, and W17, and Sh5.

[378] See above, microscene W1.

According to the labels, these rites are performed expressly for Khnumhotep II's ka. In the Egyptian conception of identity, the ka was the vital force of life within each person. This aspect of the individual was created at the same time as the body, and accompanied it through life. At death, the ka left the body, and was rejoined with the individual through the rites of burial. The ka could be immanent in images of the deceased, for example, in statues and in painted images. Offerings were made to and rituals performed for the ka.[379]

S5. Offering Bearers

The fourth and fifth registers of this wall depict processions of offering bearers. The bearers are both male and female and bear various titles associated with the running of an Egyptian estate. There is also a sealer who brings up the rear of the procession, and a scribe: the "Overseer of the House of Eternity, Niu."[380]

These offering bearers bring a variety of foods, including most of the items piled up before Khnumhotep II and his wife. The lower register begins with the inscription "Offering the cattle by the Guardians of the House of Eternity to the Great House for the ka of the Hereditary Prince, the Count, the Overseer of the Eastern Desert, Neheri's son, Khnumhotep."[381] The herdsmen at the head of this procession lead live cattle and desert animals such as an ibex, an oryx, and a gazelle. Again, these animals and foodstuffs are seen being collected, caught, or processed on the west and north walls.

Figure IV.53: Procession of Offering Bearers

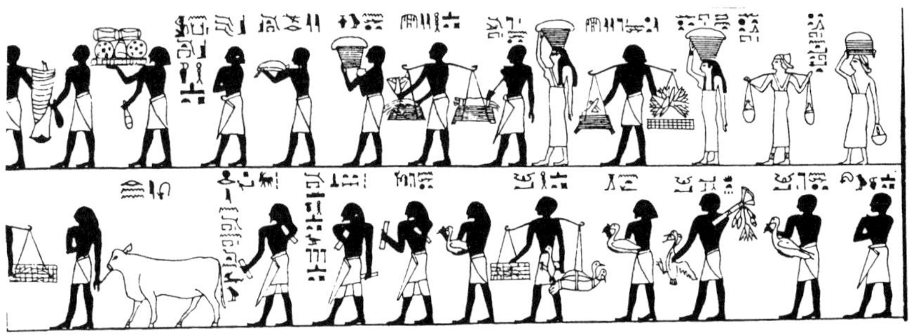

379 Kaplony, "Ka," LÄ III, col. 275-77; Wente, "Funerary Beliefs," col. 19.

380 *ỉmy-r pr n pr ḏt Nỉw.*

381 *ms ỉw3w ỉn s3w nw pr ḏt r ḥwt-ˁ3 n k3 r-pˁ ḥ3ty-ˁ ỉmy-r ḫ3swt ỉ3btt Nḫri s3 Ḫnm-ḥtp.*

Figure IV.54: Bringing Live Animals

S6. Slaughtering and Burnt Offerings

The lowest register of this wall depicts the slaughtering of animals and the preparation of burnt offerings. The second of these activities is clearly a result of the first, as a man carries a tray piled with various cuts of meat from the butchers to the officials making the offering. The animals being slaughtered cannot always be identified, but include cattle; on the basis of the animals being presented in the register above, it is likely that desert animals were sacrificed here also.[382]

This scene is very important, as it provides meat for the funerary meal, for the nourishment of Khnumhotep II and Khety. It also has ritual significance, and represents a complex ceremony. This rite begins with the "lassoing of the *nga (ngȝw),*" which perhaps makes reference to the desert hunt on the north wall, and is referred to here in vignettes which show the animals being brought to their knees and fettered. The animals are then slaughtered and butchered, their throats slit and various choice parts, such as the foreleg and the kidney fat, removed. Several of the men who perform and oversee this ceremonial slaughtering speak to one another, saying, for example: "Cut well, for the day(light) is with you!,"[383] and "I will do as you wish!"[384]

Figure IV.55: Butchering Cattle

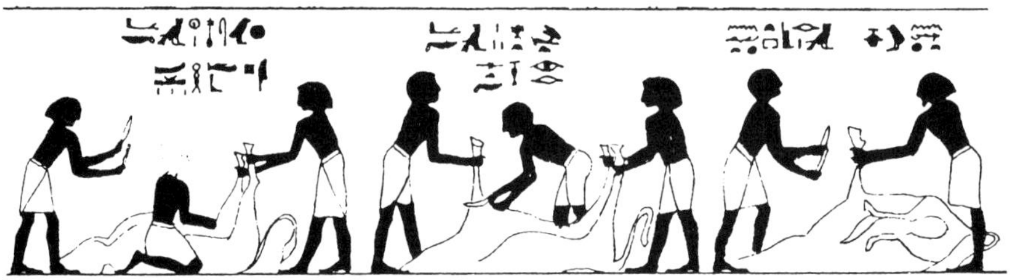

The ritual of binding and butchering sacrificial animals was an important part of the offering ceremony, and its proper performance was

382 This is explicit in Tomb 2, see Newberry, BH I, Pl. 17.

383 *ḫw.s nfr rʿ m-ʿ.k.*

384 *ir.(i) r ḥst.k.*

thought to guarantee eternal renewal.[385] The fettering of the animals is thought to have symbolized victory over the enemy, and both the butchers and their victims can be seen as representatives of various gods. For example, in a ritual text of the 12th Dynasty, the Ramesseum Dramatic Papyrus, Horus is the butcher and Thoth is the victim, while Isis supervises the slaughter.[386] Ceremonial slaughter was thought to imbue the deceased with the divine power of the animal, which was transmitted to the recipient through the burning of the offering.[387]

Figure IV.56: Burnt Offerings

S7. False Door

Khnumhotep II's false door is a small stela surmounting a niched base, carved in raised relief in the midst of the autobiography that occupies the dado. The slab stela is badly damaged, but enough remains to demonstrate that it depicted a figure of Khnumhotep II (although not labeled in any way) seated on a low-backed chair before a table of offerings, most of which has been damaged. He is in the same pose as that seen above (microscene S1), facing right, with one hand fisted on his knee and the other holding a flail to his chest.

The false door, which developed from stelae set into niches in the façade of Early Dynastic tombs, is often considered to have been the focal point of the Old Kingdom tomb chapel. It served to allow the ka of the tomb owner to pass between the burial chamber and the cult chapel, thus enabling the deceased to receive offerings.[388] Although the false door is quite prominent in several of

---

385 A. Eggebrecht, "Schlachten," LÄ V, col. 639.

386 Sethe, Dramatische Texte, pp. 109-144; Otto, "Hunting Ritual," p. 171. Otto believes that this ritual predates the canonization of the Osiris myth, accounting for the fact that it is Thoth, rather than Seth, who is the victim. He adds the observation that in later times, the bull may have symbolized Seth as the Osirian myth became more prevalent ("Hunting Ritual," p. 175).

387 Altenmüller, "Opfer," LÄ IV, col. 580.

388 Cf. Wiebach, Scheintür, p. 198; Haeny, "Scheintür," LÄ V, cols. 563ff. Wiebach summarizes the various definitions of the function of this element in Scheintür, pp. 63ff.

the earlier Beni Hasan tombs,[389] it has been reduced to an abbreviated form in Tomb 3.

Figure IV.57: False Door

Much of the original function of the false door seems to have been taken over by the expanded scene on the wall above; however, it still retains its unique function of allowing the ka of the deceased to move from the burial chamber, where it inhabited the mummy of the tomb owner, into the tomb chapel. Here the vital force of the tomb owner could be manifested in the images on the walls and in the statue in the shrine. In this way, the ka could receive the offerings presented to it, and thus render the cult effective. The placement of this false door is related to its function: it lies directly above the small libation table attached to the side of the main burial shaft, which leads to the chambers in which Khnumhotep II's body would have been interred.

OVERVIEW OF SOUTH WALL AS A WHOLE

The south wall is focused on the celebration of the funerary/offering meals of Khnumhotep II and his wife Khety. In one sense, it is a result of the actions depicted on the west and north walls, since it is concerned with the actual performance of the offering cult and thus illustrates the eventual destination of the various foodstuffs and livestock shown in the process of production or collection on those walls. The link between the north and west wall on the one hand and the south wall on the other is made explicit by the use of the term *per djet* on the west wall, the comparable north wall of Tomb 2, and the south wall here. This *per djet* seems to refer to the funerary estate, whence goods for this cult are derived.[390]

The cult is being celebrated here specifically for the kas of Khnumhotep II and Khety. The feasts in which they are partaking would have been ritually offered to them (or to their mummies or ka statues) both at their funerals, and

---

[389] See for example Tombs 29 (Newberry, BH II, Pl. 30), 17 (Newberry, BH II, Pl. 12), and 2 (Newberry, BH I, Pl. 12).

[390] See above, microscene N2.

then as invocation offerings on special feast days such as are listed around the doorway to the tomb and the shrine. Daily offerings probably would also have been made. These meals would have served to magically feed Khnumhotep and his wife for eternity.

The main burial shafts in which Khnumhotep II and Khety were buried lie at the foot of this wall. In the Old Kingdom, false doors were generally placed directly above burial shafts.[391] Here the wall as a whole has taken on much of the role of the false door while the false door itself, which is below, has been reduced to a subsidiary function; however, the false door still acts as the gateway between the burial chamber and the tomb chapel.

All of Khnumhotep II's immediate family members are depicted and carefully labeled on this wall. This careful identification serves two purposes. On the one hand, these are the people who will be performing the cult, thus their depiction magically ensures that the offerings will be made and rites performed. On the other hand, after their own deaths they will share in the offering cult represented here. Indeed, some of them may themselves have been buried in this tomb, which has space for a number of coffins in the burial chambers.

SHRINE

The decoration of the shrine is focused around the remains of a rock-cut statue at the back, which would have represented Khnumhotep II seated on what Newberry describes as a throne,[392] but which is too badly damaged to identify with any certainty. This statue is flanked by painted depictions of his chief wife Khety on the left and his mother Bakt and an unidentified woman on the right. Inscriptions in the friezes on the north and south walls request invocation offerings for Khnumhotep II's ka, and ask for offerings from the king through Osiris (north) and Anubis (south).[393]

The north and south walls of the shrine depict a cult meal celebrated by Khnumhotep II's children; the north part of the west wall portrays the female Sealer Tjat, and the south part of the west wall is too badly destroyed to decipher. The shrine is illustrated in Plate 6.

---

391 Wiebach, Scheintür, pp. 198-202.

392 Newberry, BH I, p. 53. Of the entire statue, only this "throne" remains.

393 *ḥtp di nsw di Wsir nb Ḏdw di.f prt-ḫrw n r-pʿ ḥ3ty-ʿ ḥry-tp m int ntrt sm3t m pr P3ẖt ḥ3ty-ʿ Ḫnm-ḥtp ir.n Bʿḳt.*
ḥtp di nswt di 'Inpw tpy ḏw.f di.f prt-ḫrw n k3 n im3ḫw ḥ3ty-ʿ r-pʿ ḥ3ty-ʿ ḥry-tp i3wt ḥwt-nṯr P3ẖt sw33 sr.f sw m ḳ-ḏsrt {Ḫnm-ḥtp}.

Figure IV.58: Shrine - Microscenes

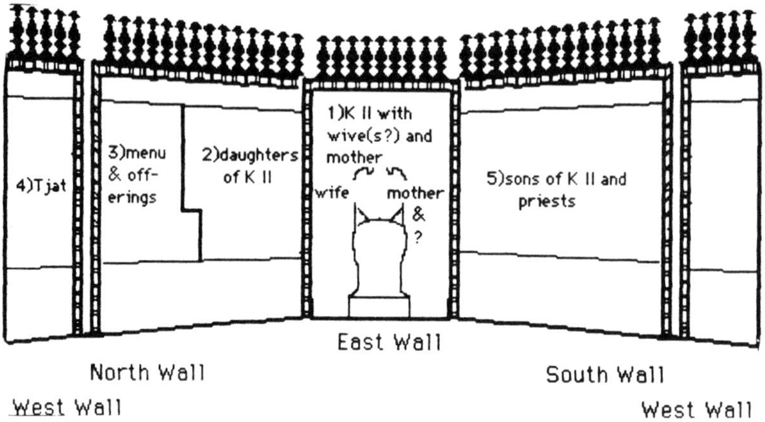

Sh1. Cult Recipients

The recipients of the cult are the rock-cut statue of Khnumhotep II and the incised and painted portraits of the three women around him. The inscriptions associated with these women are badly damaged; enough remains only to identify one as his mother and another as his chief wife, Khety.

The identity of the third woman cannot be reconstructed with certainty, but she is quite likely Khnumhotep II's grandmother. She is probably not his second wife, Tjat, as this woman appears as one of the celebrants of the cult (microscene Sh4). All of the daughters of the tomb owner who appear on the south wall are accounted for among the children paying homage to the deceased (microscene Sh2), except for Satip, daughter of Khnumhotep II and Tjat. This daughter would seem, however, to be too unimportant to be given such a prominent place in the cult. Of close female relatives, this leaves only a sister or a grandmother. Khnumhotep's maternal grandmother was the wife of Khnumhotep I, and thus a countess, she therefore seems the most likely candidate. This woman, whose name was Satip, appears, in fact, with her husband on the north wall of his tomb.[394] Her participation in her husband's cult would not have precluded her position in the tomb of her grandson, since Khnumhotep II's mother Bakt, also shown here, would presumably have participated also in her husband Neheri's cult.

It is tempting to try to associate this group of figures with mythic prototypes, so that in addition to depicting Khnumhotep II and the important women in his life, they would represent a group of divine beings. There are a number of groups of divinities who could be portrayed by such a grouping, for example, Horus with his wife Hathor, his mother Isis, and his grandmother

---

[394] See Newberry, BH I, Pl. 46.

Nut, but such identifications must be considered purely speculative, with little evidence available in this context to support them.[395]

Sh2. Daughters of Khnumhotep II

On the north wall are three daughters of Khnumhotep II and Khety, who face their father and make a gesture of respect with their right arms to their chests.[396] These women also appear on the south wall of the main chamber, where they again help to perform their parents' cult.

Figure IV.59: North and Northwest Walls of Shrine

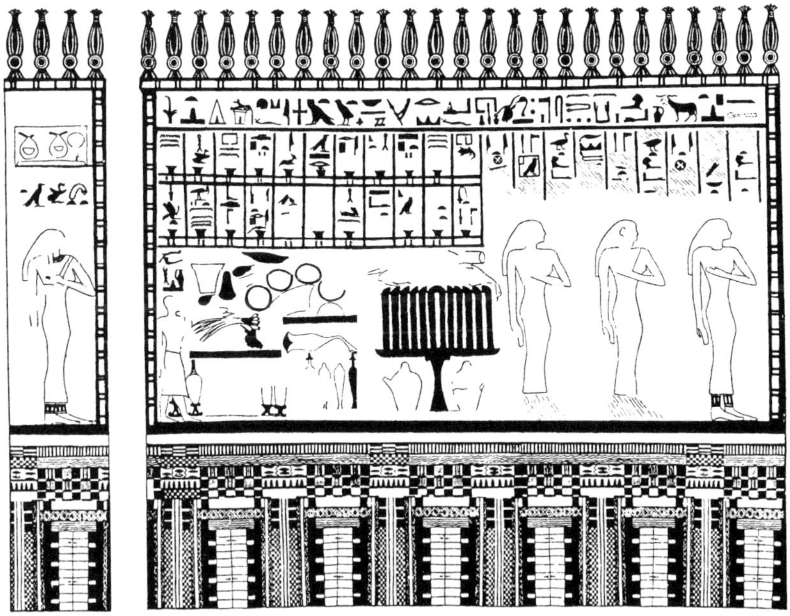

Sh3. Menu and Offerings

Behind the women is a small menu, an offering table, and a pile of offerings. This menu contains twenty-two items, including incense, natron, bread, beer, wine, and libations. The offerings piled below contain many of these same items. Shown at a smaller scale behind the offerings is yet another official named Khnumhotep. This scene can basically be considered an abbreviated version of the menu and offerings on the south wall (microscenes S1 and 2), and serves to provide for the eternal nourishment of the cult recipients.

---

395 For the various mythic beings that can be represented by groups of women, see Troy, Patterns of Queenship.

396 Cf. Brunner-Traut, "Gesten," LÄ II, cols. 575-76.

Sh4. Tjat

On the northwest wall of the shrine is a single female figure, labeled as "the Sealer, Tjat."[397] This woman also appears on the east and south walls of the main chamber; her children with Khnumhotep II are seen on the north and south walls of the main chamber, and again on the south wall of this shrine. Above Tjat's head are several bags of eye-paint. This commodity was an important cult item, and appears on the menu on the south wall of the main chamber. Its importance is underlined by microscene N4, in which eye-paint is brought to Khnumhotep II by a group of Asiatics.

Sh5. Sons of Khnumhotep II and Priests

The south wall mirrors the north, except that the women are replaced by men. At the largest scale are the sons of Khnumhotep II and Khety: Nakht, Khnumhotep, Neheri, and Netjernakht. Nakht, as on the south wall, acts as chief cult celebrant, presenting a waterfowl to his ancestors.

Figure IV.60: South Wall of Shrine

Behind Netjernakht is another son named Khnumhotep, seen on the south wall of the main chamber as the son of the tomb owner and Tjat. He is followed by Bakt and Khnumhotep, both of whom are sealers. The last figure on this wall is a lector-priest named Nakht, above whom is written the phrase: "Horus is justified!" This phrase, as discussed in several contexts within the

397 *sḏ3wtt Ṯ3t.*

main chamber, seems to stress the identification of Khnumhotep II, or perhaps his eldest son, with Horus.[398]

OVERVIEW OF SHRINE AS A WHOLE

The shrine is in some ways a reiteration of the south wall, as it depicts an abbreviated version of the offering cult, but to an expanded group of recipients. It can be seen as a development from the false door of the Old Kingdom. Such doors could incorporate a statue inside the niche; this niche later grew to the form of a small shrine such as is found here.[399] Like the south wall, this shrine forms a link between the world of the living and the *Duat*, where Khnumhotep II, Khety, Bakt, and Satip(?) would have dwelt for eternity.

---

398 See above, microscene W1.

399 Wiebach, Scheintür, pp. 142 and 198.

CHAPTER V : CONCLUSIONS
TOMB AS COSMOGRAPH

On the basis of the information presented in the previous chapter, the decoration of the tomb chapel as a whole can be seen as deliberately chosen and coherently organized to form significant patterns. The patterns created by the placement of the individual scenes help to express the ways in which the chapel functioned. Important themes are grouped together, and the walls can be read in a sequence.

Like a New Kingdom temple, this tomb chapel can be interpreted as a cosmograph, magically ensuring that the rites carried out within the microcosm of the chapel would be effective within the macrocosm of which it was a model. In this particular case, three miniature cosmoi can be identified, each graphically expressed by the choice and arrangement of the decoration within the chapel. The first cosmos is the personal cosmos of Khnumhotep II within his local realm of Beni Hasan. Within this realm, the funerary cult of the deceased noble would have been celebrated by his family, ensuring his personal survival in the Afterlife. The second cosmos is the royal cosmos, in which Khnumhotep II functions as a reflection of the king. The third cosmos is the larger cosmos of the ancient Egyptians, within which Khnumhotep II acts as a delegate of the creator god and the sun god.

The position of Khnumhotep II within each of these three cosmoi reflects Egyptian concepts of hierarchy and delegation.[1] Within the larger Egyptian cosmos, the creator god plays the central role, forming the cosmos from the limitless darkness and waters of chaos. The sun god, as a common embodiment of the creator god, reiterates this process on a daily basis, thus

---

[1] See also above, Chapter 1.

ensuring the proper continuance of the Egyptian world. The king takes on various aspects of the role of the creator/sun god in certain of his ceremonial roles, both before and after death. He fights and is victorious over the elements of chaos which intrude into the created cosmos, thus maintaining the proper order, and also, through his own death and resurrection, reiterates the cycle of creation. Both specific duties and parts of the essence of the creator/sun god are thus passed on to the king. The king can then pass some of these duties and this divine essence on to his nobles.

As discussed in Chapter IV in the context of specific scenes, identification of Khnumhotep II acting as a divinity in his own right, as the king, and through the king as the sun/creator god is made explicit in a number of the scenes contained with this tomb chapel. Images of the deceased noble are labeled as *netjer (nṯr)*, "god, divine," several times on the west wall (see microscenes W1 and W9). In many scenes, Khnumhotep II wears the ceremonial beard of royalty and divinity. He is associated several times with Osiris (see microscenes W16 and S1), and he may be specifically associated a number of times with Horus (see discussion in microscene W1). Three scenes in particular, the desert hunt and two of the marsh hunts, are taken directly from the royal repertoire, and represent Khnumhotep II in a royal pose which is in turn symbolic of the king acting as an embodiment of a god (see microscenes N1, E1, and E3). Thus Khnumhotep II's roles within these cosmoi are represented quite explicitly in this decoration.

TOMB CHAPEL 3 AS A GEOGRAPHICAL MODEL OF THE COSMOS

Several types of ancient Egyptian cult structures have been shown to function as models of the Egyptian cosmos. This cosmic modeling serves to provide an arena in which the rituals carried out will be effective, both within their cult setting and in the larger cosmos or cosmoi which these settings represent. The same type of cosmography is present within the tomb chapel of Khnumhotep II. Three cosmoi are depicted simultaneously within this arena: the personal cosmos of Khnumhotep II, represented by the landscape of Beni Hasan; the royal cosmos, represented by the landscape of the terrestrial realm; and the Egyptian cosmos as a whole, evoked in its entirety within this tomb. Thus the deceased noble is provided with settings in which he can function in each of his three roles: as the specific person Khnumhotep II; as a reflection of the king, bearing royal power which has been delegated to him; and as an incarnation of the creator/sun god, with divine power which has been passed on to him through the king.

THE PERSONAL COSMOS OF KHNUMHOTEP II

The specific environments indicated within the scenes on the walls of this chapel are arranged in such a way that they map the local environment of

Beni Hasan, within which Khnumhotep II functioned during his lifetime. Thus a personal cosmos is created, within which the deceased noble can function after death.

Figure V.1: Environments Depicted in Tomb 3

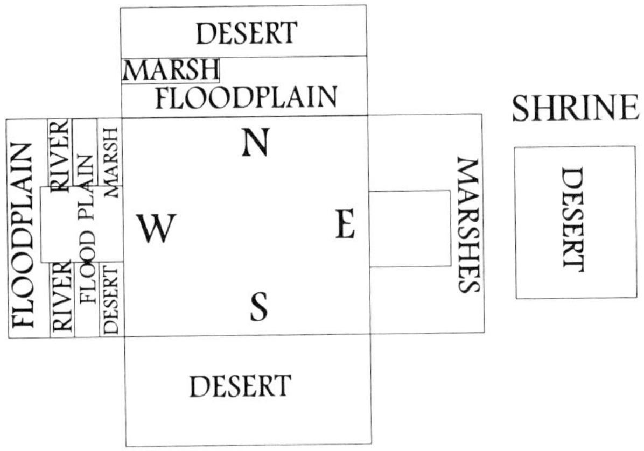

The west wall of this tomb reflects the geographical environs of Beni Hasan quite accurately, with the river in the center, surrounded by the floodplain, which in turn is bordered by marshes and canals. The eastern desert in which the tomb sits is evoked in the lowest register, and a local temple or temples, probably located in the western floodplain, can be inferred from the scene in the highest register.

Figure V.2: Map of the Local Landscape Provided in Tomb 3

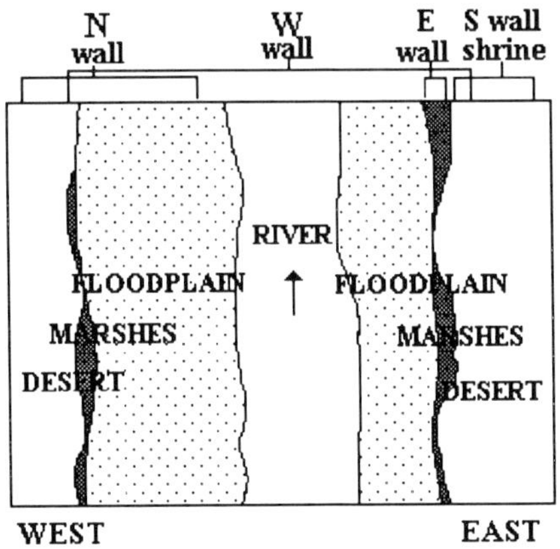

The north wall depicts the western floodplain, on which Khnumhotep II's estate would have been located, bordered by a desert landscape; as would have been the case in reality, a swampy area also intrudes between these two environments. The east wall evokes the marshes which formed in the low-lying areas at the edge of the floodplain. The south wall and the shrine both represent the locale of the tomb itself, where the offering cult would have been carried out, and thus the desert.

If these walls are rotated so that they are parallel to one another and laid side-by-side with the west wall in the center (see Figure 5.2), a clear picture of the local landscape emerges. The west wall, as described above, illustrates the entire Beni Hasan region, from the western floodplain to the eastern desert. The north wall also shows the floodplain and the desert. Khnumhotep II's agricultural estates were presumably primarily on the west bank, since there is little cultivable land on the east bank, so this area must therefore be the western floodplain and the western desert. The south wall continues the theme of the eastern desert, where the tomb was located, and the east wall and shrine reiterate this environment by evoking the cult place in the desert and depicting the marshes at its fringes.

THE ROYAL COSMOS

The royal cosmos, as the arena in which the king (or, here, Khnumhotep II as the king's delegate) is effective, is represented by the terrestrial realm. Much of the interpretation and mapping suggested for Khnumhotep II's personal cosmos also applies to this royal realm, only with an expanded scope. The landscape of the Nile Valley, with the river in the center flanked by the floodplains, deserts, and marshes, is again represented by the west, north, and south walls. The east wall can be seen here as representing, in addition to the marshes at the edge of the floodplain, the swamps found in the Nile Delta. The areas which represent the desert evoke the lands outside of Egypt, in which foreigners, embodiments of the forces of chaos, dwell. This is shown literally on the north wall, where actual foreigners are depicted in conjunction with the desert.

THE EGYPTIAN COSMOS

The scenes within this tomb chapel can also be interpreted as evocations of the most important elements of the Egyptian cosmos as a whole: the terrestrial realm; the sky (the celestial realm and the domain of the sun god); and the hazardous realm of the *Duat*, through which beings must pass in order to be reborn, found both below (as the Netherworld, the territory of Osiris) and above the earth. The earthly realm is the most literally depicted; the realms of the sky and the *Duat* are evoked more indirectly.

The terrestrial realm is represented, as discussed in the descriptions of the local and royal cosmoi, on all of the walls, but primarily on the west and north walls. All of the key geographical elements of the earth are depicted here: river, floodplain, marshes, and desert/foreign lands. The desert, where the tomb is located, is also implied by the funerary meal depicted on the south wall.

The celestial realm is evoked in a number of ways within this chapel. The sun, and therefore the sky, is brought into the tomb by the use of a sunshade to protect the tomb owner in a scene on the west wall (see microscene W4), and the appearance of sunshades on one of the boats in the pilgrimage scene (microscene W16). On the north wall, the celestial realm as the territory of the sun god is implied through the numerous solar symbols contained within the desert hunt scene (microscene N1).

Several other tombs of this period, from Deir el-Bersheh, evoke the celestial realm in a way similar to that found in numerous royal and divine contexts, namely by the use of stars on the ceiling of their chapels.[2] Neither this technique, nor the other relatively clear means of evoking the sky, such as the use of *pt* (sky) signs framing scenes or processions of sun disks across the ceiling[3] are found in Khnumhotep II's tomb chapel. It is significant, however, that the ceiling of this chapel is decorated with representations of textiles, echoing the roof of a tent shrine.[4] In actuality, this tomb is cut into the rock of a cliff face, and the sky lies far above the rock-cut vaults of the ceiling. In contrast to this reality, the textile pattern painted here suggests that this ceiling is meant to be thought of as a roof, with the sky just beyond. In fact, in Tomb 2 at Beni Hasan, a similar pattern includes areas that have been interpreted as skylights, flaps that would have been opened to let in the sun.[5] Thus the sky is brought into the tomb indirectly, through the choice of ceiling decoration. The evocation of a roof implies a sky from which the cult structure is protected, transforming and eliminating the mountain which in reality lies between the chapel and the sky.

The territory of the sun god, both above and below the earth, is also evoked in a more symbolic fashion by the pilgrimage scene (microscene W16). This scene can be interpreted on one level as a representation of the daily journey of the sun god, who sails across the sky during the day and rows through the *Duat* at night.

The netherworld aspect of the *Duat* is also evoked by a number of specific areas of the chapel. On the west wall, it is implied by the scene which depicts the building of the tomb (microscene W9). The south wall of the chapel represents the world of the tomb, located in reality in the desert, but implies the netherworld aspect of the *Duat* into which the burial chambers, located directly below this wall and linked to it by the false door, symbolically reach.

---

[2] See Silverman, Bersheh Reports I (forthcoming).

[3] Finnestad, Image, pp. 11-13.

[4] See above, Chapter II.C.

[5] Cf. Badawy, Architecture II, p. 136.

In fact, based on other coffins of this period from Beni Hasan,[6] it is quite likely that the coffins within these burial chambers would themselves have been decorated with more specific representations of the *Duat,* through which the deceased had to travel in order to be reborn. Khnumhotep II can be identified on this wall with Osiris, the ruler of the *Duat.*

Figure V.3: Map of Larger Cosmos Evoked in Tomb 3

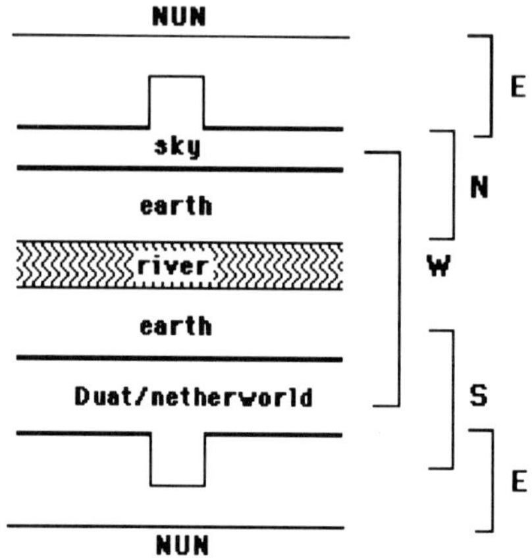

The east wall and shrine together represent the landscape of Creation. The act of Creation is accompanied by the suppression of disorder, represented here by the scenes on the east wall, where Khnumhotep II is shown acting as the creator/sun god (and as the king), imposing order on the chaotic elements of the marshes.

The four walls and the shrine can be arranged to form a map of the cosmos as a whole. In the center is the west wall, illustrating the earth with the sky implied above and the netherworld evoked below. Flanking the west wall are the north wall, reiterating earth and implying the celestial realm, and the south wall, evoking the *Duat* below the earth. Bordering the north and south walls is the east wall/shrine complex, which shows the landscape of Creation which emerges from the primeval chaos of the Nun.

TEMPORAL CYCLES WITHIN THE THREE COSMOI

As discussed in Chapter I, the ancient Egyptians perceived two principal types of time. Linear (*djet*) time was associated with Osiris, and was thought of as timelessness, in which the cosmos had a finite beginning and

[6] See Willems, Chests of Life, pp. 66-67 and 234ff.

end, and remained unchanging between these fixed points. Cyclical (*neheh*) time was the province of the sun god, and focused on "eternal recurrence," expressed by the perpetual re-enactment of the process of Creation seen daily in the rising and setting of the sun and yearly in the changing of the seasons.[7] The Egyptians also recognized historical time, which has both linear and cyclical aspects.

All of these types of time are represented in this tomb chapel.[8] Historical time is illustrated by references to the life of Khnumhotep II on earth, which includes royal dates both in the autobiography and in association with the procession of Asiatics on the north wall (microscene N4). This historical time is primarily the province of the tomb owner's personal cosmos. Also contained within this personal cosmos is the cycle of the life, transfiguration, rebirth, and Afterlife of Khnumhotep II.

Figure V.4: Temporal Cycles represented in Tomb 3

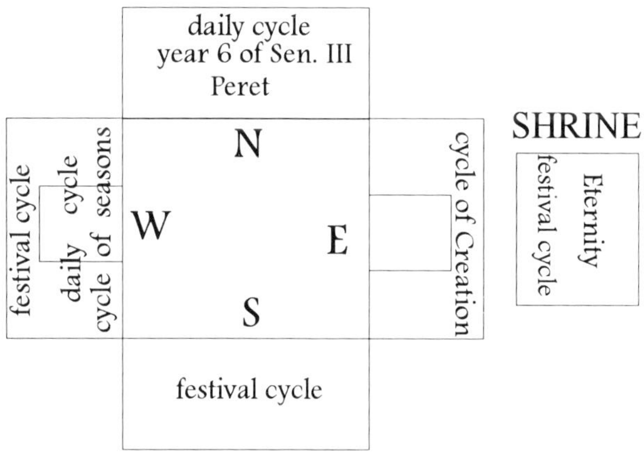

The solar cycle permeates all three of the cosmoi present within the tomb chapel. This cycle can be seen within the terrestrial realm, and also as a reflection in the cosmic realm of the death and regeneration of the tomb owner. The seasonal cycle of the Egyptian calendar is clearly illustrated on the west wall, where reference is made to the seasons of *Akhet*, *Peret* and *Shemu*. This seasonal cycle is generated by the solar cycle, and represents the appropriate action of *neheh* in the terrestrial realm. The cycle of festivals appears explicitly in several lists of festivals (around the entrances to the tomb and the shrine, and in the biography), and implicitly in the statue transport scene on the west wall, and the cult celebrations on the south wall and in the shrine.

[7] Allen, Genesis in Egypt, pp. 31ff.; Wente, "Funerary Beliefs," p. 22.

[8] These types of time, although very differently expressed, are also represented in the royal ka chapel of Nephepetre at Dendereh, see O'Connor, "Nebhepetre's Chapel."

Linear time is represented in this tomb by the perpetual funerary cult on the south wall (and in the burial chambers), where Khnumhotep II is associated with Osiris. This expression of linear time spills from the personal cosmos of the tomb owner into the royal and larger Egyptian cosmoi. Linear time is also evoked by the east wall/shrine complex, which can be seen as an illustration of *sep tepy (sp tpy),* the first time of Creation. This Creation is re-enacted daily, and is thus also an expression of cyclical time.

The thematic elements of these temporal cycles seem to be arranged quite deliberately, in fairly consistent spatial groupings, so that there seems to be an underlying narrative sequence contained within this chapel. This sequence moves from the north wall, counterclockwise around the walls of the main chamber and into the shrine.

Figure V.5: Narrative Progression in Tomb 3

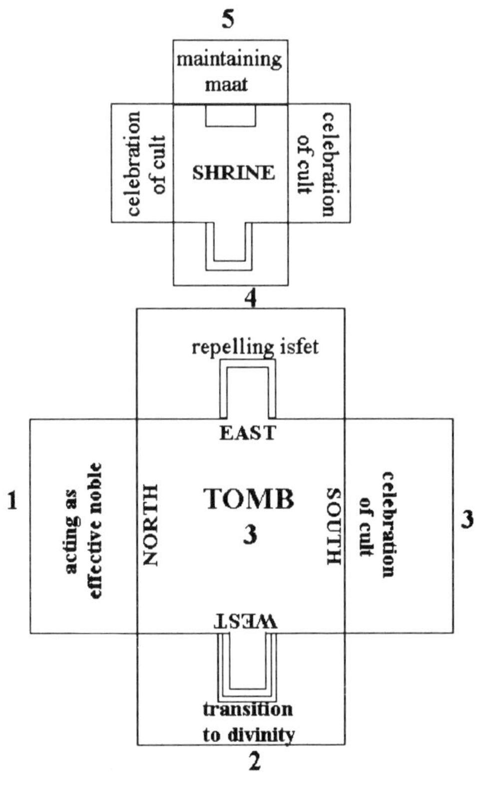

The north wall begins with Khnumhotep II, within his personal cosmos, collecting "taxes" for the king, and thus symbolizes the appropriate execution of the proper duties of an effective noble. This wall includes a specifically dated event, namely the arrival of a group of Asiatics in Year 6 of Senwosret II, providing a distinct historical event firmly based in the real world. The king is represented on this wall through the agency of two royal scribes, one of whom records the herds brought before him and the other of whom hands Khnumhotep a papyrus recording the arrival of the Asiatics. Also within this personal cosmos, this wall refers to the setting up of the mortuary estate, granted to him by the king and the gods (a concept expressed by the *hetep di nesu* formulae) and thus acts as ensurance that necessary goods, specifically desert and domesticated animals, birds, and eye-paint, will be available for Khnumhotep II's funerary cult.[9]

At the same time, the actions on this wall can be interpreted as actions of Khnumhotep II as a delegate of the king, who receives both foreign and domestic tribute. Both the royal and the Egyptian cosmos are evoked by the desert hunt scene, in which Khnumhotep acts as the king, who is in turn acting as the creator/sun god, repelling the forces of chaos. The tribute scene is also expressive of the maintenance of the power of Egypt over her foreign foes, who themselves represent the forces of chaos.

The next stage of the narrative is shown on the west wall, which depicts preparations for and the celebration of the funeral (represented by the pilgrimage to Abydos), and thereby illustrates and ensures Khnumhotep II's effective transformation to a state of divinity, both in his own right and through identification with Osiris. Also on this wall is the procession of a cult statue, including associated rituals, which would presumably have occurred during festivals, possibly before Khnumhotep II's death, but certainly afterwards, in connection with his mortuary cult.[10]

The south wall is dedicated to a primary consequence of the activities on the north and west walls, namely the successful celebration of the funerary meal and its associated rituals, which served to purify and nourish the deceased within each of these cosmoi for eternity. A false door on this wall briefly reiterates the theme of the cult celebration and provides a link with the burial chambers below. The flail held by Khnumhotep at the cult meal and on the false door helps to associate him within the royal cosmos as the king, the maintainer of *maat* (the proper order of the universe), and within the larger cosmos with Osiris as prototypic king and ruler of the *Duat*.[11]

The east wall shows Khnumhotep II hunting in the marshes; he is seen here as one who repels *isfet*, the forces of chaos. Khnumhotep II is enabled to do this by the proper celebration of the cult shown on the south wall; conversely, by performing this task, he perhaps ensures that he will be considered worthy of the cult,[12] and guarantees that the cosmic order necessary for the performance of the cult will be maintained. These scenes clearly

---

9 See above, Chapter IV, Overview of North Wall.

10 See above, microscene W1.

11 See above, microscene S1.

12 Cf. Barocas, "Décoration," pp. 434-35.

associate him in the royal cosmos as playing the role of the king, and in the Egyptian cosmos as playing the role of the creator/sun god. The position of these scenes flanking the doorway to the shrine also serves an apotropaic function, similar to the purpose served by images of the king smiting his foreign enemies found on the pylons of New Kingdom temples.[13] Around the doorway to the shrine, the cult is represented textually through a list of festivals and offering prayers to Anubis and Osiris.

The decoration inside the shrine again depicts the celebration of the cult, this time for images of Khnumhotep II and his wife, mother, and grandmother(?) by his male and female children and his second wife, Tjat. This celebration, like the meal on the south wall, is a result of the appropriate completion of the actions depicted on the north and west walls, and is also performed in consequence of the successful maintenance of the cosmic order represented on the east wall. The east wall/shrine complex then begins the cycle again, by representing the re-enactment of the Creation.

The autobiography which runs around the dado of the main chamber, although it does not correspond spatially with the themes on the walls above it, reflects the same basic themes and thus fits reasonably well into the sequence outlined here. It begins on the north side of the east wall and, like the narrative sequence, runs counterclockwise around the walls of the main chamber, ending at the entrance to the shrine. Like the pictorial decoration, the autobiography stresses the themes of appropriate behavior of the tomb owner towards his ancestors, his king, and his dependents. It again reflects themes of hierarchy and delegation, and underlines his role in helping to maintain the cosmic order, as a proxy for the king on one level, and as an embodiment of the creator/sun god on another.[14]

ORIENTATION: KHNUMHOTEP II WITHIN THE THREE COSMOI

Analysis of the patterns of the flow of action and the orientation of major figures within these three cosmoi produces significant results. As discussed above, the basic direction of the flow of action in this chapel is from west to east around the walls of the main chamber and to the back of the shrine. In general, it can be said that the majority of the figures performing the cult on the north and south walls face east; on the east and west walls, such figures tend to face the central axis of the tomb. Images of the recipient of the cult, on the other hand, face either west or away from the axis of the tomb.[15]

---

[13] See above, Overview of East Wall.

[14] Cf. Altenmüller, "Lebenszeit," pp. 83-84.

[15] Cf. Davies and Gardiner, Amenemhēt, p. 18. On this subject, see also Kanawati, in "Living and dead," pp. 219-21.

The ultimate recipients of the actions portrayed in the chapel are the cult figures within the shrine: Khnumhotep II with his wife Khety, his mother Bakt, and possibly his grandmother Satip.[16] In addition to the rock-cut cult statue of the tomb owner, there are a number of two-dimensional figures of Khnumhotep II in the main chamber which also act as cult images and receive the benefit of the actions performed. On the north wall of the main chamber, a large-scale standing figure of the deceased faces west and dominates the wall; a comparable seated figure appears directly opposite on the south wall. A smaller-scale figure of Khnumhotep II also appears on the west wall, facing away from the central axis and thus receiving the action (microscene W4).

Figure V.6: Orientations of Major Figures and Flow Pattern of Tomb 3

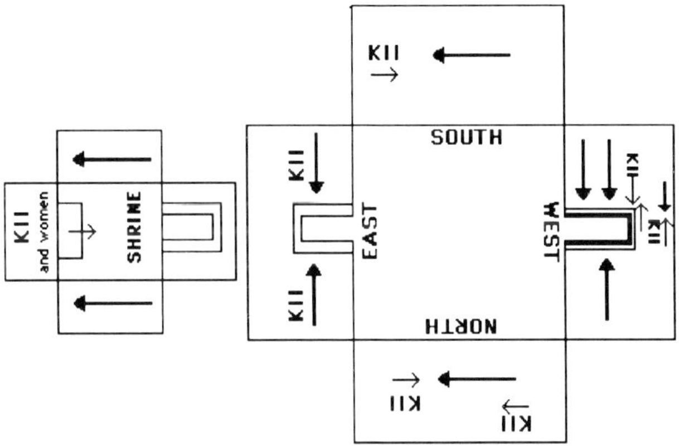

To complicate and at the same time clarify the situation, several figures of Khnumhotep II are portrayed facing inward (east or toward the axis of the tomb) and thus, according to this interpretation, are contributing to the action rather than (or in addition to) receiving its benefits. These figures appear on three of the four walls of the main chamber: on the north wall, the tomb owner is shown hunting in the desert, on the west wall, he is actively supervising the manufacturing, and on the east wall he is shown (three times) hunting in the marshes. The distinction between the inward- and outward-facing figures of the tomb owner correlates with his pose: when he faces in, he is active; when he faces out, he is passive. The inscriptions associated with these figures also support this interpretation: the labels on the north and south walls describe actions being performed for Khnumhotep II, while the labels on the east wall

[16] The reasons for the addition of wife and ancestresses are not entirely clear, but may be connected in some way with the concepts of regeneration and eternal rebirth. See above, microscene Sh1.

describe actions being performed by Khnumhotep II.[17] This circumstance invites comparison with the decoration of gods' temples, where the cult celebrant (the king), who is active on the part of the god, faces in, while the cult recipient (the god) faces out.[18]

In this chapel, therefore, Khnumhotep II both receives and generates the action, reflecting the actions of both the "king" (actor) and the "god" (beneficiary and enabler). In fact the three primary instances where Khnumhotep II faces inward and is actor rather than recipient mirror images that are seen in both earlier and later royal iconography in royal and/or gods' temple; in these contexts, they are interpreted as symbolic of the role of the king as the repeller of the forces of chaos.[19]

Figure V.7: Khnumhotep II Within the Larger Cosmos

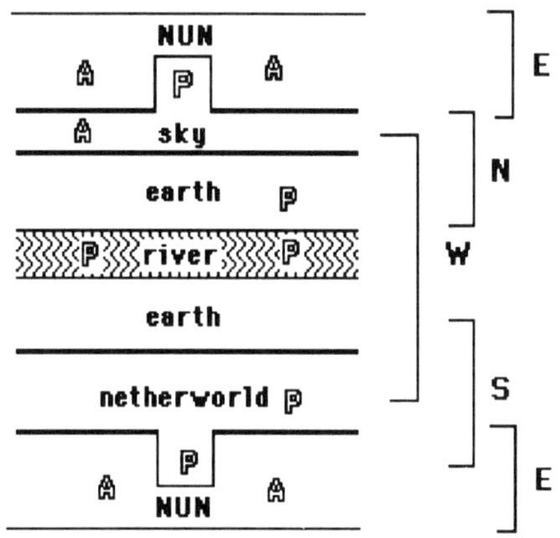

Placing the figures of Khnumhotep II which appear on these walls in their approximate places within the maps of the three cosmoi illustrated within the chapel reinforces the significance of their orientation on all three levels of interpretation. Within the bounds of the local cosmos, Khnumhotep II is shown as passive, receiving the goods and services which are offered to him, and as active, performing acts which help to ensure the order of his personal cosmos. On the national level, he acts as a proxy for the king, receiving goods and services in return for his actions in repelling *isfet* and maintaining the cosmic order. Within the larger Egyptian cosmos, Khnumhotep II is passive inside the bounds of the earth, where, as a god, he receives homage. He is also passive in the netherworld aspect of the *Duat* represented by the funerary cult.

---

17 See above, microscenes N2, S1, E1, and E3.

18 Arnold, Wandrelief, p. 128.

19 See, for example, discussion of microscene E1.

However, in the implied celestial realm and in the area of the primeval mound (on the east wall) he is active, paralleling the actions of a king within a temple, who in turn acts as an embodiment of the creator god.

TOMB CHAPEL 3 AS "PRIVATE MORTUARY TEMPLE"

A number of useful parallels can be drawn between this tomb chapel and gods' temples, to which it is analogous in a number of ways. As discussed above, the monuments best suited to this discussion are gods' temples of the New Kingdom.[20] There are also important similarities between this tomb chapel and royal mortuary temples, as well as royal ka chapels;[21] these monuments provide an important link between the private tomb chapel and the god's temple, since royal mortuary temples and royal ka chapels also provide loci of worship for beings who have both earthly and divine natures, beings who are, like the deceased noble, identified after death with divinities.[22]

The architectural correspondence between the chapel of Khnumhotep II and the god's temple of the New Kingdom provides a basis for this analogy.[23] Comparisons of the decoration of a typical god's temple with the decoration of this tomb chapel reveals a functional concordance between the two types of monument. Finally, since the tomb, like the temple, can be viewed as a cosmograph, the rites performed within the tomb, like the ritual associated with a temple, would have been rendered effective in the framework of the cosmos or cosmoi represented.

CONCORDANCE WITH TEMPLE DECORATION

An essential function of the decoration of the temple of the New Kingdom and later (seen also in at least one Middle Kingdom royal ka chapel) was to act as a link between the earthly and divine realms by mirroring the Egyptian cosmos, and thus providing a setting in which the rituals carried out would be effective.[24] The architectural resemblance between Khnumhotep II's tomb chapel and the basic plan of the New Kingdom temple has already been noted.[25] This resemblance, coupled with the fact that this chapel also mirrors the cosmos, and seen in the light of the many ways in which the tomb owner symbolically plays the role of a king or god, suggests that comparison of these types of structures would be a useful exercise.

---

[20] See above, Chapter I.B.

[21] O'Connor, "Nephepetre's Chapel," and Simpson, "Kenotaph," LÄ III, cols. 387-391.

[22] Cf. Blumenthal, "Königsideologie," LÄ III, cols. 528-529; Frankfort, Kingship and the Gods, pp. 148-51.

[23] Cf. Badawy, Architecture II, p. 195.

[24] Cf. Gundlach, "Tempelrelief," LÄ VI, cols. 407-411.

[25] Cf. Badawy, Architecture II, p. 195.

As discussed above, the decoration of a god's temple focuses on the mutually beneficial relationship between the king and the gods.[26] The king, as chief celebrant, acts to empower and placate the god through various rituals and offerings and is given in return the wherewithal to rule in the earthly realm. The rule of the king is then represented in a number of ways, with emphasis on his role as guarantor of the cosmic order as maintainer of the cosmic order (*maat*) and repeller of the forces of chaos (*isfet*).

Figure V.8: (=1.6): Flow of Benefits in the God's Temple

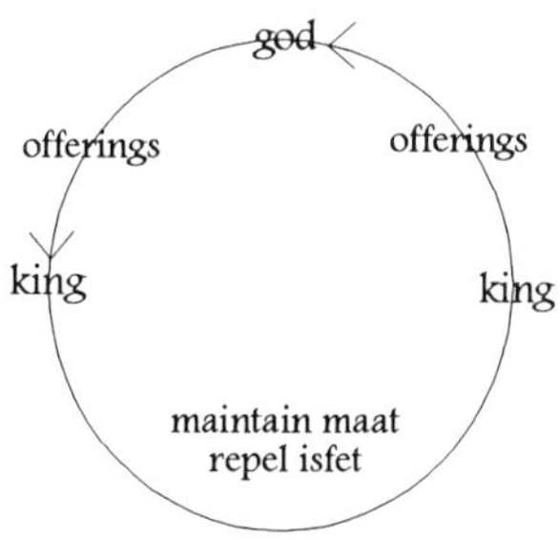

Temple decoration can be divided into the following broad categories: 1) rituals performed by the king in order to empower and placate the god and his or her image, which include scenes of the king presenting various libations, symbols, and items such as clothing and oil to the god; 2) the cult meal, presented by the king to the god, which also serves to empower and placate the divine being; 3) the emergence of the cult image from the temple in festival processions; 4) the presentation by the god of symbols of kingship and power to the king and the assumption by the king of his rule on earth; 5) and the performance by the king of his role on earth, expressed through images of him performing as both maintainer of *maat* (e.g., seated on his throne holding the insignia of his power) and repeller of *isfet* (e.g., as victor, primarily against foreign enemies).

The system within Khnumhotep II's tomb chapel is more complex than that found in temples for a number of reasons. First, kings and gods are mentioned in textual references, but are not represented visually within the tomb, probably for reasons of decorum.[27] Second, the primary relationship in the tomb chapel is between the deceased noble and the descendants and officials who celebrate his cult.

---

[26] See above, Chapter I.B.

[27] Baines, "Temple Symbolism," p. 14.

Figure V.9: Flow of Benefits in the Tomb Chapel

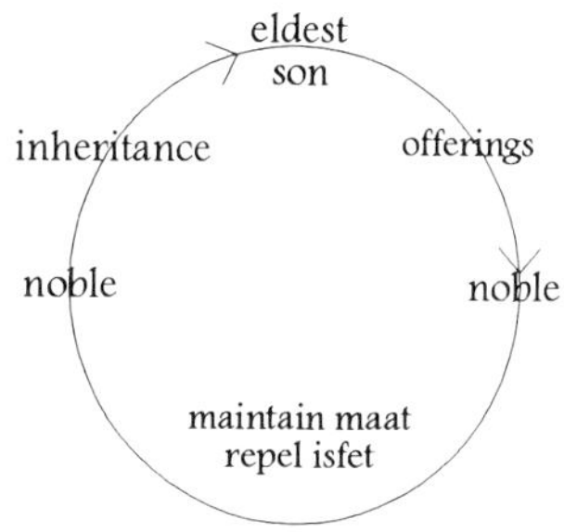

Despite these differences, functional equivalents for many of the categories of scenes found in gods' temples can be found in Tomb 3. Some of the scenes in this tomb chapel are very similar to their counterparts in temples, while others are quite different, but serve analogous purposes. The role of king as placator of the divine being is represented in the tomb by the priests of Khnumhotep II, primary among whom is his eldest son. The role of the noble can be seen as a mirror of the roles both of the god who receives the benefits of the cult and of the king who acts to guarantee the cosmic order in the earthly realm. The direction of the flow of benefits is reversed from the temple to the tomb chapel; the offerings come from the eldest son, and in return, he is permitted to inherit his father's position.[28]

Rendering the Image Effective

Gods, although they die and are reborn within the cycle of the cosmos, are always gods, and need only enter their cult images to become effective on earth. Nobles, like individual ruling kings, must die, be properly buried, and follow the correct path through the dangers of the Hereafter before they can gain whatever form or forms of divinity are appropriate to their status. Scenes in temples which show the presentation to divine images of clothing, libations, and the like, serve to animate these images by preparing them for the presence of the god whom they represent. The functional equivalent of these scenes in Khnumhotep II's chapel are scenes relating to the funeral: depiction of the proper celebration of these rites would ensure that the deceased would be able to function properly within his tomb chapel.

There are a number of scenes found in Tomb 3 which are connected to the funeral. These include representations of the manufacture of burial equipment, the production and storage of food for the funerary meal (and

28 The situation in Khnumhotep II's chapel is complicated slightly by the fact that he has two "eldest" sons, his true eldest son, Nakht, who was destined to inherit political power in another nome (biography, lines 121-50), and his second eldest son, Khnumhotep, who was his political heir in the Oryx nome (biography, lines 151-60). It is the true eldest son, Nakht, who acts as his father's chief cult celebrant.

subsequent cult meals, and perhaps also for the statue cult), and the pilgrimage to Abydos. These scenes are placed together on the west wall, and illustrate the manufacturing of the items needed for Khnumhotep II's cult, including his tomb (microscenes W2-6 and 8-9), the agricultural and hunting activities which will provide food for the funeral as well as for daily and festival offerings (microscenes 7, 11-13, and 15); the storage of the equipment and grain (microscene W10); and the husbandry of cattle (microscene W14). This theme is reiterated on the north wall, in the gathering together of materials for the funerary cult.

Paired riverine voyages, which can be interpreted as depicting either a real or symbolic pilgrimage to Abydos which may have formed part of the actual funeral or an actual ritualized drama, are also found on the west wall (microscene W16). This scene represents the actual transition of the tomb owner to his incarnation as a fully empowered deceased being: Khnumhotep II travels to Abydos as a mummy and then returns, having "learned the requirements," as an effective spirit (probably to be thought of as an *akh*, although this is not stated here).[29] The deceased noble is identified with Osiris through his mummiform shape and royal beard; he (or, alternatively, his eldest son) is associated with Horus in the inscription "Horus is justified."[30] Khnumhotep II also emphasizes the importance of a properly built tomb in his autobiography.[31]

Celebration of Cult Meal

In this category of scene, there is a one-to-one correlation between gods' temples and this tomb chapel. The basic layout of such scenes is the same in both types of monument. The cult recipient (god or noble) is shown seated before a table heaped with offerings, primarily of food. Rites performed by the chief cult celebrant (king for the god, eldest son for the noble) and other priests are often seen as elements of these scenes. Offerings are brought by bearers, and sacrificial animals are slaughtered. Menus list the offerings made and rites performed.

In Tomb 3, there are two depictions of this scene: one on the south wall; and a second in the shrine. Reference is also made to this theme on the false door. The most complete elaboration of this theme is found on the south wall, on which both the tomb owner and his wife are shown seated before offering tables. On this wall, various rites of purification are performed, and spells are recited, sacrificial animals are slaughtered and burnt offerings made, a wide variety of food and drink is presented to Khnumhotep II and Khety, and rites, food, and equipment are enumerated on large offering list. The chief cult celebrant is the eldest son of the noble; this son was traditionally the heir of his

---

29 See above, microscene W16.

30 See above, microscene W1.

31 Lines 170ff.

father and was thus analogous to the king in a temple, who was shown as the heir of the god. In the shrine, a less extensive illustration of a cult meal focuses on the rock-cut statue of the tomb owner flanked by his mother and grandmother(?) and his wife; these images can be compared to the cult statues of gods found within the shrines of temples.[32] As is the case in temple contexts, these rituals serve both to maintain and placate the deceased for eternity.

Festival Processions

A type of scene frequently found in temples is the illustration of a festival procession, in which the cult image would have been taken from its shrine and carried to visit other temples. In Tomb 3, the statue-transport scene (microscene W1) fulfils this function. This scene shows a statue of the tomb owner within a portable shrine which is dragged on a sledge; associated inscriptions read: "the god will come that the earth might be protected," and "open the doors of heaven that the god might come forth." It is likely that this statue was taken on processions from its cult place (in a ka chapel associated with the temple or in the tomb complex) during certain festivals,[33] just as the image of a god traveled from his temple on certain occasions.

Assumption by King/Noble of Proper Role

In a temple, the pharaoh is handed the symbols of kingship by the god, and is often shown in celebrations related to his ascension to the throne, such as the coronation. In these scenes, the king represents both a specific king, such as Ramesses II, and an embodiment of the concept of kingship, a prototypic king who was awarded the throne in the mythological realm.

The theme of the assumption of earthly power is represented in Tomb 3 by scenes and texts which relate to the awarding of titles and privileges to Khnumhotep II by the king. This theme is found explicitly in the autobiography, which states the specific circumstances under which he gained his position within the nome. It is also implicit in the scenes on the north wall of the chapel, which show him supervising the gathering of *inw*, which serves both as taxes for the king and material for the mortuary estate, under the auspices of the king. Explicit reference to specific kings is made in this tomb in the autobiography (Amenemhēt II) and on the north wall (Senwosret II).

---

[32] There are no free-standing examples of statues preserved from the Pharaonic Period, probably because they would have been made of precious materials (see for example, the shrines containing statues from the tomb of Tutankhamun, e.g., Brown, et. al., Treasures of Tutankhamun, #38, or several later examples of sculptures which may be cult statues, cf. Reeves and Taylor, Howard Carter, ppp. 112 and 172). However, several rock-cut temples have rock-cut statues preserved within their shrines (e.g., Abu Simbel, see Porter and Moss, Topographical Bibliography VII, pp. 95-119).

[33] See above, microscene W1.

King/Noble as Guarantor of Cosmic Order

The role of guarantor of cosmic order is carried out in temple contexts by the king, acting as surrogate on earth for the gods, and, in particular, for the sun god.[34] There are two aspects to this role: the king must maintain *maat*, the proper cosmic order, and repel *isfet*, the forces of chaos.[35] His role as maintainer of *maat* is symbolized by depictions where he is seated on a throne, holding the symbols of his office. As repeller of *isfet*, he takes an active pose, symbolically holding the forces of chaos at bay through victory over Egypt's enemies. Scenes that illustrate this theme are icons of the king smiting his enemies, battle scenes, and hunting scenes.

These themes are also illustrated in Tomb 3, in scenes where Khnumhotep II can be associated with the king on one level and identified with Osiris on another. As maintainer of *maat*, Khnumhotep II twice appears seated and holding a flail, once at the celebration of the cult meal (microscene S1), and again on the slab stela of the false door (microscene S7). The royal flail is the symbol of the coercive power of the king, and can be interpreted as identifying its holder (including Osiris) as playing the role of the maintainer of *maat*.

There are also several small-scale seated figures of the tomb owner which function as determinatives on the exterior jambs of the entrance doorway and on the exterior jambs of the doorway to the shrine, and thus are close to the large-scale figures of Khnumhotep II as the repeller of *isfet* on the east wall. These figures can be usefully compared to seated statues of the king which are placed in front of the pylons to a temple, in close conjunction with images showing the king triumphing over foreign enemies. The cult statue of Khnumhotep II may also have shared in this iconography, representing the noble as maintainer of the cosmic order, but all that can be concluded with certainty from what little remains of this statue is that it was seated.

Khnumhotep II is also shown quite explicitly as the repeller of *isfet*. The primary expression of this theme is on the east wall, where three versions of hunting in the marshes appear. Each of these images shows the noble subduing the chaotic forces of the marshes, symbolized by either fish or waterfowl. The association of the noble with the king is these scenes is enhanced here by the fact that he wears a kilt similar to the royal *shendyt* and the ceremonial beard.[36] On the north wall of this tomb, this theme is again expressed in the image of Khnumhotep II hunting in the desert, again a region of disorder; this theme also appears in royal contexts.[37] These scenes are analogous to temple scenes showing the king victorious over his enemies, and

---

[34] Frankfort, Kingship and the Gods, pp. 239 and 148-51.

[35] Cf. Assmann, "State and Religion," p. 58.

[36] See above, Overview of East Wall.

[37] See Borchardt, *Sa³ hu-re* II, Pl. 17.

like them, can be found found in close conjunction to images which identify the tomb owner as maintainer of *maat*.

BENI HASAN TOMB 2

It is also possible in the context of this study to demonstrate briefly that the tomb chapel of Khnumhotep II does not represent an isolated example of deliberate cosmography. Tomb 2, which directly antedates Tomb 3, also displays an attempt at cosmological mapping, and shares many of the features of Tomb 3. A discussion of this second tomb must of necessity be summary; many of the conclusions reached in the context of Tomb 2 are based on the more thorough analysis done for Tomb 3, and are thus not fully developed in this presentation.

There is a great deal of correlation between the tomb chapel of Tomb 2 and the chapel of Tomb 3. To begin with two tombs are of the same architectural type, and are almost identical in plan. Many of the same themes are represented in both tombs, and there are important parallels between certain of the corresponding walls from one chapel to the other, most particularly the north and south walls. However, there are important differences between the two tombs, in terms of scenes chosen, details within similar scenes, and placement of various themes. The most crucial difference between the two tombs is that while in Tomb 3, Khnumhotep is clearly identified as playing the role of the king on the one hand and of the sun/creator god on the other, neither of these identifications are shown explicitly in the chapel of Tomb 2.

Figure V.10: Plan of Tomb 2

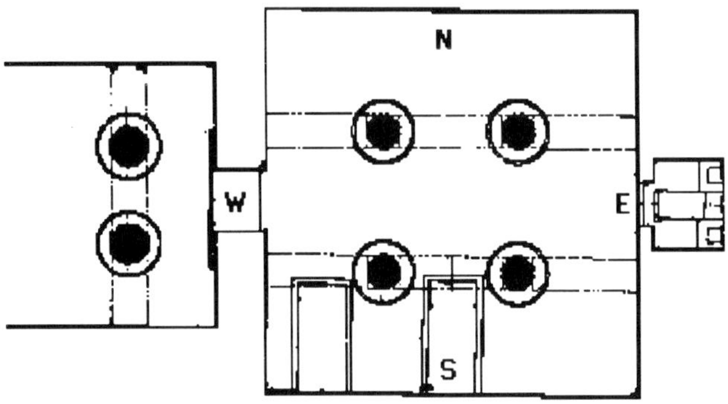

BRIEF DESCRIPTION

Tomb 2 was built during the reign of Senwosret I for a Great Chief of the Oryx Nome called Ameni.[38] It was cut into the northern end of the tomb-lined cliff face at Beni Hasan, just north of Tomb 3. Like Tomb 3, it is approached from the valley via a path lined with boulders that runs up the scree slope fronting the cliff into an open courtyard. The layout of this tomb is very similar to that of Tomb 3; it too is of the "portico-in-antis" type, with a columned portico set before the entrance doorway, a square main chamber, and an axially set shrine.

The area around the entrance doorway is carved with a number of titles and epithets of Ameni, and prayers to Anubis and Osiris. A long autobiography was inscribed on the thickness of the doorway into the main chamber of the tomb. The main chamber is decorated with scenes and texts; below the scenes is a rough dado, and on the ceiling, vaulted like that of Tomb 3, are patterns which resemble textiles, including elaborate squares which are thought to mimic skylights.[39] Two rectangular burial shafts with slightly raised edges were cut into the floor in front of the south wall. Neither shaft was cleared by Newberry, thus their interior plans, dimensions, and contents are unknown.[40]

A shrine, about 2.2 meters east to west and 2.7 meters north to south is reached by a doorway in the center of the east wall. Its floor-level is several inches higher than the floor of the main chamber, and its ceiling is about 3 meters lower. Pivot holes on the threshold of this shrine indicate that double doors once separated the two chambers.[41] In the center of the back wall was carved a seated statue of Ameni, now destroyed, flanked by standing statues of his wife Hetepet and his mother Henu.

A khekher frieze above a band of colored rectangles runs around the tops of the paintings in both the main chamber and the shrine; vertical borders of colored rectangles and "rope" patterns frame each wall. Below the scenes are from two to six variously colored (gray, red, white, yellow, black, or blue) horizontal bands, and below these bands is an unpainted dado, 0.76 meters high, which was left somewhat rough.

There are many different activities, mostly related to production of both perishable and non-perishable items for the funerary/divine cult of Ameni and his wife Hetepet depicted on the west wall, which is divided into two halves by the doorway that forms the entrance to the tomb. To the south of the door

---

[38] Ameni's autobiography is dated to Year 43 of Senwosret I, which is said to be "equal to year 25 in the Oryx nome of ...Ameny..." Thus Ameni seems to have ruled the Oryx nome from Year 18 to Year 43 of Senwosret I. (See Lichtheim, Autobiographies, pp. 137-39.)

[39] Badawy, Architecture II, p. 136. For illustration see Newberry, BH I, Pl. 6.

[40] Newberry, BH II, p. 72.

[41] Newberry, BH I, p. 20 and Pl. 4.

are various stages in the gathering and processing of food,[42] all of which, with the exception of fish, would have been offered both at the funeral and at the perpetual celebration of the cult. Also to this side of the entrance is a false door and related scenes. The right side of the wall, from the third register to the eighth, depicts the primary elements of the invocation offering: bread, beer, fowl, and cattle. This half of the wall focuses on Ameni and his wife Hetepet, who are depicted together twice: once in the marshes, and again on the false door. It is significant that they are both present, and shown as equal in scale to one another. This may indicate Hetepet's importance,[43] and may add stress to themes of sexuality and rebirth.[44] To the north of the doorway are scenes of manufacture and agriculture.[45] All of the activities shown here would have taken place on the noble's estate on the floodplain, either in workshops or in the cultivated fields. The materials produced in both the manufacturing and agricultural activities would have served as provisions and equipment for the burial in particular and the funerary cult in general.

The north wall illustrates three major themes: hunting in the desert, statue transport,[46] and the collection of taxes/setting up of the estate. The hunting scene is in some ways a continuation of the themes current on the west wall, as it also shows the gathering of "raw materials" in the form of wild desert animals for the cult. It is also an expression of the continual victory of order over chaos. The statue transport scene shows a statue of Ameni, identified as divine, being installed in its cult place and moved at festivals. The single large-scale figure of Ameni on this wall supervises the collection of taxes for the king (alluded to in the presence of royal scribes), which can be seen simultaneously as goods for his funerary estate.[47]

The east wall depicts wrestling and battle scenes and pilgrimages to Abydos and Busiris.[48] The wrestling and battle scenes present the theme of conflict and chaos; these scenes may serve an apotropaic function, as they surround and thus perhaps protect the entrance to the shrine. It is odd that no one watches the wrestling, and neither side clearly wins the battle. It may be the case that when the shrine doors were closed, chaos prevailed, but when they were opened, the statue of Ameni that appeared between them supervised these activities and thus symbolically created order within them. The

---

[42] See Newberry, BH I, Pl. 12.

[43] On size and relationships of figures, see Munro, "Größverhältnisse," LÄ II, cols. 903-906.

[44] It might also suggest identifications with Osiris and Isis in the celebration of a hypothetical "Sacred Marriage," such as is frequently alluded to by Kessler (e.g., "Jagdszene," p. 38); however, there is no conclusive evidence for such an identification.

[45] See Newberry, BH I, Pl. 11.

[46] The placement of this scene on the north wall, in contrast to the placement of a similar scene on the west wall of Tomb 3, may be due in part to the tradition, seen in Tombs 15 and 17, of placing such a scene on the north wall.

[47] Cf. above, Overview of North Wall.

[48] See Newberry, BH I, Pls. 14-16.

pilgrimage scenes, representing the transition of Ameni to a state of apotheosis, are arranged so that the boats actually move into the shrine, where Ameni's cult statue is located.

The south wall is devoted to the celebration of cult meals for Ameni and Hetepet.[49] Such celebrations would have taken place at the time of their funerals and thereafter at important festivals; by representing the ritual and offerings involved, the cult would have been magically performed in perpetuity. Ameni, like Khnumhotep II holds a flail, which associates him with both Osiris and the king as the maintainer of *maat*.[50] The burial shafts in which Ameni and Hetepet were presumably interred lie directly beneath their respective celebrations, providing direct access between the burial chambers and the cult. The decoration of the shrine reiterates the theme of the celebration of the cult meal.[51] The beneficiaries of this celebration are the rock-cut statues of Ameni, his wife Hetepet, and his mother Henu.

Table V.1: Significant Differences between Tombs 2 and 3

Tomb 3	Tomb 2
Khnumhotep II shown active, as repeller of *isfet*, in desert and marsh hunts.	Ameni shown only passive. Desert hunt performed by nameless hunters.
Khnumhotep II fishing and fowling in marshes (east wall).	Ameni is not shown fishing or fowling in the marshes.
Statue-transport scene on west wall.	Statue-transport scene on north wall.
Pilgrimage (to and from Abydos) scene on west wall.	Pilgrimage (to Abydos and Busiris) scene on east wall.
False door on south wall; dedicated to Khnumhotep II alone.	False door on west wall, more elaborate than false door in Tomb 3, and dedicated to both Ameni and Hetepet.
Autobiography on dado below scenes.	Autobiography on thickness of entrance doorway.
Shipbuilding shown (west wall).	No shipbuilding shown.
No wrestling or battle scenes.	Wrestling and battle scenes on the east wall.

49 See Newberry, BH I, Pls. 17 and 18.

50 See above, microscene S1.

51 See Newberry, BH I, 19 and 20.

SYNTHETIC ANALYSIS OF THE TOMB CHAPEL OF AMENI

The type of analysis that was carried out for the tomb chapel of Khnumhotep II can also be applied to this chapel. As was the case for Tomb 3, the decoration here was carefully arranged and reflects certain similar patterns. Ameni's chapel can also be interpreted as a model of the cosmos, and can also be usefully compared to a New Kingdom temple. There are important differences between the two chapels which make Tomb 2 more difficult to analyze. In fact, certain aspects of this second chapel can only be understood through reference to Tomb 3.

Organization of Themes and "Narrative" Flow

The organization of themes in Tomb 2 is in many ways comparable to the arrangement found in Tomb 3, since much of the decoration in the two tombs is extremely similar. The differences between the two chapels, however, affect the narrative flow: instead of moving from north, around the tomb counter-clockwise and into the shrine, the flow moves from the north wall to the west wall, then to the east wall, and back again to the south wall and into the shrine.

The north wall shows the tomb owner acting as an effective noble on behalf of the king and at the same time represents the setting up of the funerary cult and perhaps also the statue cult; this is, as was the case in Tomb 3, the first phase of the narrative. The second phase moves to the west wall, where the production of food and manufactured items for the funeral and cult is illustrated. The third stage is also anticipated on this wall, since some of which items are presented to Ameni and Hetepet directly, and a cult celebration in the form of a false door is represented on this wall.

The second stage continues on the east wall, where the pilgrimage scenes (to both Abydos and Busiris) are located. The offering cult seen on the false door is reiterated and expanded on the south wall, and is again for both Ameni and Hetepet. The cult meal is shown a third time within the shrine, this time with the addition of Ameni's mother as a recipient of the offerings. Thus the narrative sequence jumps between the various walls to a certain extent, and is not as clearly designed as it was in the chapel of Khnumhotep II. These differences may reflect the influence of earlier tombs on the chapel of Ameni.

Figure V.11: Organization of Themes in Tomb 2

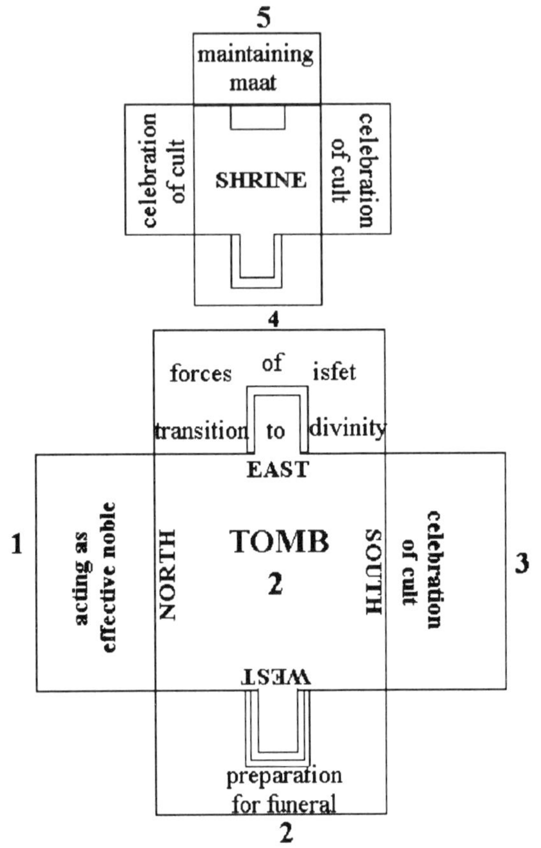

TOMB 2 AS A MODEL OF THE COSMOS

Ameni's chapel can also be seen as a model of the cosmos at several levels. It mirrors the local and national Egyptian landscape, and at the same time acts as microcosm of the Egyptian universe as a whole. The same basic elements of the Egyptian landscape present in Tomb 3 are again present in Tomb 2, and can be interpreted in a similar, although not identical, fashion.

The west wall can be interpreted as including the basic elements of the Egyptian landscape, namely river, floodplain, marsh, and desert. These elements are not arranged as coherently as was the case in Khnumhotep II's chapel. The river is not actually depicted at all, but can be reconstructed from the watery environments of the seining and clapnetting scenes, normally interpreted as marshes, but here perhaps seen as riverine, that fall in the middle of the wall. The north wall again shows the desert and the floodplain. The east wall depicts the river and an area that might be a fortress on the floodplain or in the desert where a battle took place, or might be somewhere outside of Egypt. The shrine and the south wall both evoke the desert.

Figure V.12: Elements of the Landscape Present in Tomb 2

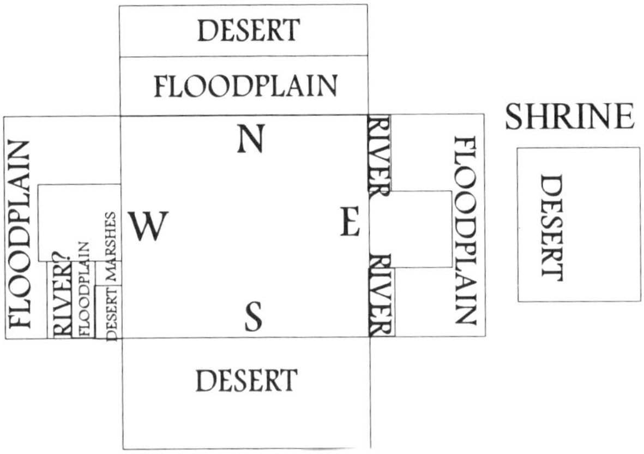

If the four walls are laid side by side, an image of the Egyptian landscape can be ascertained. This image is somewhat difficult to construe, in comparison to the situation in Tomb 3. The layout is not as neat or clear, and must be reconstructed carefully.

Figure V.13: Model of The Egyptian Landscape in Tomb 2

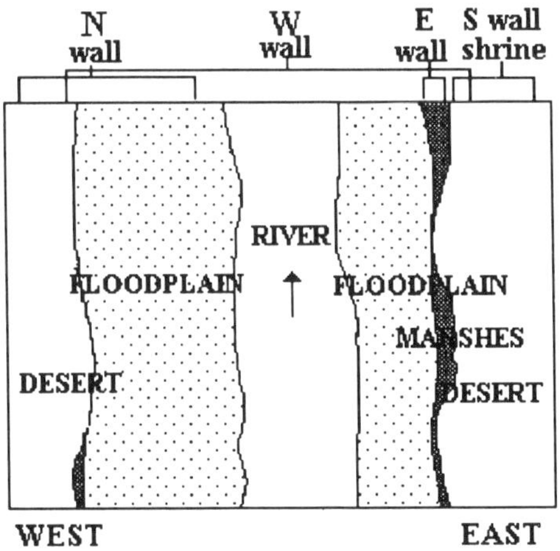

The east wall is in the center, illustrating the river and what is guessed here to be the eastern floodplain. To one side (the east) of this is the shrine and south wall, both representing the eastern desert, in which the tomb was located. The north wall illustrates the area to the west of the river, namely the western floodplain, on which Ameni's estate was presumably located, and the western desert, in which the hunt may have taken place. The west wall spans the landscape from the western floodplain (north of the door and south of the door, upper registers), through the river (marsh scenes), to the eastern floodplain, marshes, and desert.

The larger Egyptian cosmos is also evoked in this tomb. The west wall represents the real Egyptian landscape that forms the center of this cosmos. The south wall reflects the netherworld, where the deceased and his wife dwell in the Afterlife. The north wall depicts the earthly world bordered by chaos: the desert can be seen as the border between Earth and Nun, where the sun travels. The river represented on the east wall reiterates this watery border area, and the wrestling and battle scenes represent Nun itself, an area of chaos. The shrine with its images of the tomb owner and his mother and wife, by analogy with the temple shrine, represents the primeval mound filled with the regenerating power of the creator god; the shrine thus emerges from the chaotic surroundings of the east wall.

Figure V.14: Model of the Cosmos in Tomb 2

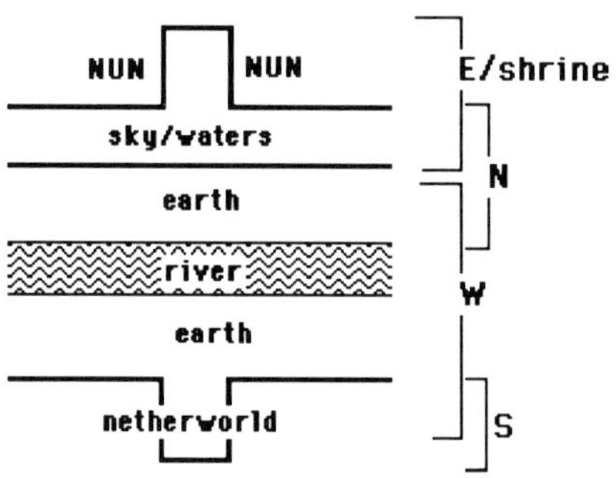

As in Khnumhotep II's chapel, the principal flow of action in this chapel is from west to east around the walls of the main chamber and to the back of the shrine. The figures performing the cult or bringing offerings face east or toward the central axis of the tomb, while the cult recipients (Ameni alone on the north wall, Ameni and his wife Hetepet on the west and south walls, and Ameni, Hetepet, and Ameni's mother Henu in the shrine) face either west or away from the axis of the tomb.

In contrast to Tomb 3, there are no active figures of the tomb owner in this chapel. Thus an important difference between the two tombs is that, whereas in Tomb 3, Khnumhotep II is shown overtly acting as the repeller of *isfet*, Ameni plays this role only covertly, if at all. There is one place in the chapel where Ameni faces toward, rather than away from, the central axis of the tomb; this is on the slab stela of the false door on the west wall. On this stela, Ameni and Hetepet are represented facing one another, thus the direction in which Ameni faces is probably due to a desire for symmetry.

Figure V.15: Flow of Action in Tomb 2

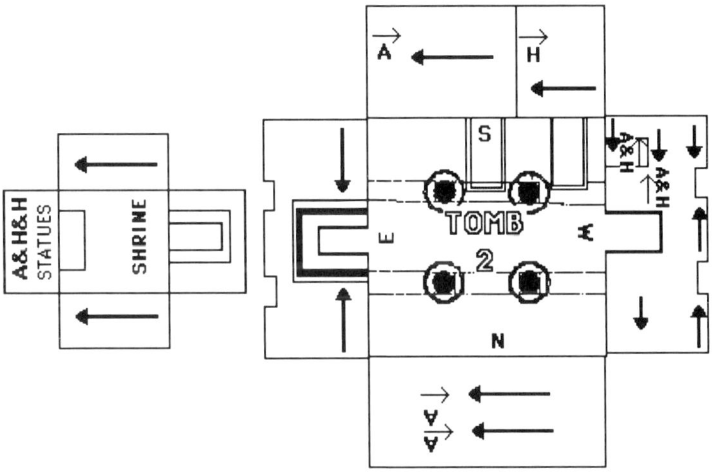

Elements of cosmogony and cosmic process are also present within this tomb chapel, but again not as clearly illustrated as they were in Tomb 3. The cosmogony can again be seen in the shrine/east wall complex, where the statue of Ameni "emerges" from the surrounding chaos. All of the various seasons of the Egyptian calendar are represented, and the daily cycle of the sun is referred to by the pilgrimage scene on the east wall and perhaps also in the hunting scene on the north wall. Linear eternity is evoked on the south wall and in the shrine.

TOMB 2 AS A "PRIVATE MORTUARY TEMPLE"

Since the architecture of this tomb is so similar to that of Tomb 3, the architectural comparison between tomb chapel and New Kingdom temple that was made for Tomb 3 can also be said to hold true for Tomb 2. The courtyard can be compared to the the festival court, the main chamber is equivalent to the hypostyle hall, and the shrine plays the role of the sanctuary.

Concordance with Temple Decoration

Since so many of the themes found in Tomb 3 are also found in Tomb 2, it is not surprising that the decoration of this chapel also can be usefully compared to the decoration of a New Kingdom temple. However, there are aspects to this comparison that must use Tomb 3 as a mediator, since they are not completely apparent from Tomb 2 alone.

The theme of the transition to divinity of some sort is represented in Tomb 2 by the same type of scene which played this role in Tomb 3: the pilgrimage scene. In Tomb 2, the river voyages shown flanking the doorway to the shrine can be viewed as conflations of pilgrimages to and from the Osirian sites of Abydos to the south and Busiris to the north; at the same time, they represent the funerary voyage. Any and all of these journeys could have been made in reality, magically through the medium of these scenes, or symbolically within the confines of the cemetery. Taken together, these scenes also suggest links with solar symbolism: the boats to one side of the doorway can be compared to the day barque of Re in which he travels across the sky; the boats to the other could represent the sun god's voyage through the netherworld at night.[52]

The celebration of the cult meal is depicted three times: on the false door, on the south wall, and within the shrine. It is interesting to note the status of Ameni's wife Hetepet as revealed by these scenes: she is shown as equal to her husband on the false door, and has her own cult, complete with menu, offering bearers, priests, and butchers, on the south wall.

The festival procession is represented here again by a statue transport scene. This particular scene has more of the flavor of a statue installation, since the statue in its shrine is met by a small group of female offering bearers. The statue here is explicitly identified as Ameni himself, and the label "the god will come that the earth might be protected" endows the image with divinity.[53]

Ameni is portrayed as an effective noble both on the north wall, where he is shown collecting taxes for the king, and in his autobiography, where he states: "I delivered all their dues to the king's house, and there was no shortage against me in any bureau of his..."[54]

In Tomb 2, Ameni is not shown overtly as the guarantor of cosmic order, except in the label to the statue transport scene mentioned above, where his image comes "to protect the earth." He is not shown hunting in the marshes, and in the hunt on the north wall, it is nameless hunters who attack the desert animals. On the east wall, neither the wrestling nor battle scenes have clear victors, thus the message inherent here might even be thought to be "Chaos prevails." However, on the north wall, the large standing figure of Ameni who receives the offerings brought to him can be seen as dominating the desert

---

52 See discussion for Tomb 3, microscene W16.

53 See discussion for Tomb 3, microscene W1.

54 Lichtheim, Autobiographies, p. 139.

hunt, and thus participating in it indirectly. The scenes on the east wall are surrounded by images of Ameni: the two large-scale images on the north and south walls, and the rock-cut statue in the shrine. Thus Ameni acts more covertly as the guarantor of the cosmic order; this aspect is there, but has not been developed the way it has in Khnumhotep II's chapel.

CONCLUSIONS

The chapel of Beni Hasan Tomb 3 can be interpreted as a small mortuary temple, comparable in function to a royal or divine temple, but on a local rather than national level. The decoration of this chapel represents the execution of a complex plan that can be read at multiple levels.

The following chart represents the three principal levels at which Khnumhotep II's tomb chapel can be read: the local level of the Beni Hasan area (the personal cosmos of Khnumhotep II), the national level of Egypt (the royal cosmos), and the universal level of the larger Egyptian cosmos. Each of these levels is represented by both temporal and spatial elements.

Table V.2: The Three Nested Cosmoi in the Chapel of Khnumhotep II

COSMOS	ROLE OF NOBLE	TIME	SPACE
Personal cosmos of Khnumhotep II	Khnumhotep II as local god in his own right	Earthly and Eternal Life of Khnumhotep II Yearly Cycle of Seasons	Landscape of Beni Hasan
Royal cosmos	Khnumhotep II as King	Yearly Cycle of Seasons Daily Cycle of Sun	Landscape of Nile Valley
Cosmos	Khnumhotep II as Sun/Creator God	Cosmogony and Cosmic Process Daily Cycle of Sun	Cosmic Landscape

Thus, in the same set of scenes and texts, three interrelated narratives function simultaneously through the delegation of authority and function from one level of the cosmic hierarchy to another. On one level, the decoration tells of Khnumhotep II's life on earth, his funeral, and his life in the Hereafter within his own personal cosmos, in which he functions as a divine being. This same narrative can be read again as the life cycle of Egypt, complete with all the features of the Egyptian landscape and all the seasons of the Egyptian year; Khnumhotep II acts as the king within this royal cosmos. The third level illustrates the life of the larger Cosmos, complete with the reiterated cosmogony, wherein Khnumhotep II plays the role of the sun god and therefore the creator god.

Both the architecture and decoration of this chapel demonstrate important analogies to both royal and divine temples of the New Kingdom. In both appearance and function, this chapel can be considered a sort of temple, in this case a private mortuary temple. Although the way in which the cosmos is evoked in this chapel is somewhat different from the method used in a temple, the ultimate effect is the same: the monument is transformed into an arena in which the rituals carried out will be effective in the universe.

Within this structure, a cult would have been celebrated for Khnumhotep II, although how long this cult may have lasted after his death cannot be ascertained. It is most likely that, in fact, this cult was effective only within the borders of the Oryx nome, and probably served primarily for the deceased noble's family. Through Khnumhotep II's symbolic identification as a reflection of both king and god, he could be seen as representing a divine being, perhaps capable of acting on behalf of those who celebrated his cult in the divine realm. In this way, his cult can be seen as an elaboration and extension of the cult of the divinized dead apparent, for example, in Letters to the Dead.

It should also be noted again that the conclusions put forth for Tomb 3 at Beni Hasan do not automatically apply to other Egyptian tombs. The analysis carried out here is very specific, and based primarily on the information found in this particular tomb. Examination of the tomb chapel of Ameni at Beni Hasan suggests that similar conclusions might be made for this monument, but a number of the most striking aspects of Tomb 3 are represented only covertly in Tomb 2, and can only be ascertained by making reference to Tomb 3. It may well be that other tombs share the temple-like function seen here, but these must be examined on a case by case basis.

The tomb chapel of Khnumhotep II thus joins other types of cultic structures which have been shown to function as cosmographs, in which the rituals celebrated become effective in the larger universe of the ancient Egyptians. This chapel represents a triumph of planning and execution, and can be read as a complex statement about its owner and his place in the cosmos.

BIBLIOGRAPHY

Aldred, Cyril. Middle Kingdom Art in Egypt, 2300-1590 B.C. London: 1950.

_____. "Grabdekoration," in Lexikon der Ägyptologie II, 853-857.

_____. "Grabrelief," in Lexikon der Ägyptologie II, 866-870.

Allam, Schafik. Beiträge zum Hathorkult (bis zum Ende des Mittleren Reiches). Munich and Berlin: 1963.

Allen, James P. Genesis in Egypt: the Philosophy of Ancient Egyptian Creation Accounts. New Haven: 1988.

_____. "Funerary Texts and their Meaning," in Mummies and Magic. (Boston: 1988).

_____. "The Cosmology of the Pyramid Texts," in Religion and Philosophy in Ancient Egypt, (Yale Egyptological Studies 3), W.K. Simpson, ed. New Haven: 1989.

Allen, T.G. Review of Helmut Wilsdorf, Ringkampf im alten Ägypten (Würzburg-Aumühle: 1939), in American Journal of Archaeology 44 (1940), 257.

_____. The Book of the Dead or Going Forth by Day. Chicago: 1974.

Alliot, M. "Les Rites de la Chasse au filet, aux temples de Karnak, d'Edfou, et d'Esneh." Revue d'Égyptologie 5 (1946), 57-118.

Altenmüller, Hartwig. Darstellungen der Jagd im Alten Ägypten. Berlin: 1967.

_____. "Eine neue Deutung der Zeremonie des *init-rd*," in Journal of Egyptian Archaeology 57 (1971), 146-153.

_____. "Zur Frage der *Mww*," in Studien zur altägyptischen Kultur 2 (1975), 1-37.

_____. "Lebenszeit und Unsterblichkeit in den Darstellungen der Gräber des Alten Reichs," in 5000 Jahre Ägypten, J. Assmann and G. Burkard, eds. Nusslock: 1983.

_____. "Abydosfahrt," in Lexikon der Ägyptologie I, 42-47.

_____. "Fabeltiere," in Lexikon der Ägyptologie II, 74-75.

_____. "Jagd," in Lexikon der Ägyptologie III, 221-224.

_____. "Jagddarstellungen," in Lexikon der Ägyptologie III, 224-230.

_____. "Opfer," in Lexikon der Ägyptologie IV, 579-584.

Anderson, J. and W. E. de Winton. Zoology of Egypt: Mammalia. London: 1902.

Anthes, Rudolf. "The Original Meaning of Mꜣʿ-ḫrw," in Journal of Near Eastern Studies 13 (1954), 21-51.

Arnold, Dieter. Wandrelief und Raumfunktion in ägyptischen Tempeln des Neuen Reiches. Berlin: 1962.

_____. Der Tempel des Königs Mentuhotep von Deir el-Bahari. Volume I: Architektur und Deutung; Vol. II: Die Wandreliefs des Sanktuares. Mainz: 1974.

_____. Gräber des Alten und Mittleren Reiches in El-Tarif. Mainz: 1976.

_____. "Vom Pyramidenbezirk zum 'Haus für Millionen Jahre,'" in Mitteilungen des deutschen archäologischen Instituts Kairo 34 (1978), 1-8.

_____ "Grab," in Lexikon der Ägyptologie II, 826-837.

_____. "Pyramiden, MR und später," in Lexikon der Ägyptologie IV, 1263-1272.

_____. "Totentempel I, II, III," in Lexikon der Ägyptologie VI, 694-711.

_____. Building in Egypt: Pharaonic Stone Masonry. Oxford: 1991.

_____, and J. Settgast, "Vorberichte über die vom deutschen archäologischen Institut Kairo im Assasif unternommenen Arbeiten," in Mitteilungen des deutschen archäologischen Instituts Kairo 20 (1965), 47-61.

Arnold, Dorothea. "Wandbild und Scherbenbefund: zur Töpfertechnik der alten Ägypter vom Beginn der pharaonischen Zeit bis zu den Hyksos," Mitteilungen des deutschen archäologischen Instituts Kairo 32 (1976), 1-34.

Assmann, Jan. "State and Religion in the New Kingdom," in Religion and Philosophy in Ancient Egypt, (Yale Egyptological Studies 3), W.K. Simpson, ed. New Haven: 1989.

_____. "Death and Initiation in the Funerary Religion of Ancient Egypt," in Religion and Philosophy in Ancient Egypt, (Yale Egyptological Studies 3), W.K. Simpson, ed. New Haven: 1989.

_____. "Schöpfung," in Lexikon der Ägyptologie VI, 677-685.

Badawy, Alexander. "The Ideology of the Superstructure of the Mastaba Tomb in Egypt," in Journal of Near Eastern Studies 15 (1956), 180-183.

_____. A History of Egyptian Architecture, Volumes 1 and 2. Giza, Berkeley and Los Angeles: 1954 and 1966.

_____. "Illusionism in Egyptian Architecture," in Studies in Honor of John A. Wilson. Chicago: 1969 (Studies in Ancient Oriental Civilisation 35), 15-23.

_____. "Festunganlage," in Lexikon der Ägyptologie II, 194-203.

Baines, John. "Temple Symbolism," in Royal Anthropological Institute Newsletter 15 (1976), 10-15.

_____. Fecundity Figures. Warminster: 1985.

_____, and Jaromir Málek. Atlas of Ancient Egypt. New York: 1985.

_____. "Restricted Knowledge, Hierarchy, and Decorum: Modern Perceptions and Ancient Institutions," in Journal of the American Research Center in Egypt 27 (1990), 1-23.

_____. "Society, morality, and religious practice," in Religion in Ancient Egypt, Byron E. Shafer, ed. Ithaca: 1991.

_____. "Kingship, Definition of Culture, and Legitimation," in Ancient Egyptian Kingship, D. Silverman and D. O'Connor, eds. (Leiden, New York: 1995), 3-47.

Balcz, H. "Die Gefässdarstellungen des Alten Reiches," in Mitteilungen des deutschen archäologischen Instituts Kairo 4 (1933), 18-36.

_____. "Zu den Szenen der Jagdfahrten im Papyrusdickicht," in Zeitschrift für ägyptische Sprache und Altertumskunde 75 (1935), 32-38.

Barguet, P. Le temple d'Amon-Rê à Karnak. Cairo: 1962.

Barocas, Claudio. "La Décoration des chapelles funéraires égyptiennes," in La mort, les morts dans les sociétés anciennes, Gnoli, G. and J.P. Vernant, eds., Cambridge: 1982, 429-440.

Barta, Winfried. Aufbau und Bedeutung der altägyptischen Opferformel. (Ägyptologische Forschungen 34.) Glückstadt: 1868.

_____. "Bemerkungen zur Chronologie der 6. bis 11. Dynastie," in Zeitschrift für ägyptische Sprache und Altertumskunde 108 (1981), 23-33.

_____. "Opferliste," Lexikon der Ägyptologie IV, 586-589.

_____. "Schakal," in Lexikon der Ägyptologie V, 526-528.

Bates, Oric. "Ancient Egyptian Fishing," in Varia Africana I (1917), 199-271.

von Beckerath, Jürgen. "Die Dynastie der Herakleopoliten (9./10. Dynastie)," in Zeitschrift für ägyptische Sprache und Altertumskunde 93 (1966), 13-20.

_____. "Kalender," in Lexikon der Ägyptologie III, 297-299.

_____. Handbuch der ägyptischen Königsnamen. Munich: 1984.

Behrmann, Almuth. Das Nilpferd in der Vorstellungswelt der Alten Ägypter. Frankfurt am Main: 1989.

Bennett, J.C. "Growth of the *Ḥtp-di-nsw* formula in the Middle Kingdom," in Journal of Egyptian Archaeology 27 (1941), 77-82.

Bidoli, Dino. Die Sprüche der Fangnetze in den altägyptischen Sargtexten. Glückstadt: 1976.

Blackman, Aylward M. The Rock Tombs of Meir, volumes I, II, III, and VI. London: 1914-1953.

_____. "The Ka-house and the serdab," in Journal of Egyptian Archaeology (1916), 250-254.

Bleiberg, Edward. "The King's Privy Purse during the New Kingdom," in Journal of the American Research Center in Egypt 21 (1984), 155-167.

Blumenthal, Elke. "Die Datierung der *Nḥri*-Graffiti von Hatnub. Zur Stellung der ägyptischen Gaufürsten im frühen Mittleren Reich," in Altorient. Forschungen 4 (1976), 35-62.

_____. "Königsideologie," in Lexikon der Ägyptologie III, 526-531.

Bochi, Patricia A. Agricultural Scenes in the Private Tombs of the 18th Dynasty: A Study in Iconographic Polyvalence. Ph.D. dissertation: University of Pennsylvania; 1992.

Boessneck, J. Die Tierwelt des Alten Ägypten. Munich: 1988.

Borchardt, Ludwig. Die ägyptishe Planzensäule. Berlin: 1897.

_____. Das Grabdenkmal des Königs *Sa3 ḥu-reʿ*, Volume II. Leipzig: 1913.

Boulenger, G.A . The fishes of the Nile. London: 1907.

Breasted, J. H. Ancient Records of Egypt, Vol. I: The First to the Seventeenth Dynasties. Chicago: 1906.

Brewer, Douglas J. and Renee F. Friedman. Fish and Fishing in ancient Egypt. Warminster, England: 1989.

Brinks, Jürgen. Die Entwicklung der königlichen Grabanlagen des Alten Reiches. (Hildesheimer Ägyptologische Beiträge 10.) Hildesheim: 1979.

_____. "Die Sedfestanlagen der Pyramidentempel," in Chronique d'Égypte 56, no. 111 (1981), 5-14.

Brovarski, Edward. "The Doors of Heaven," in Orientalia 46 (1977), 107-115.

_____. "Serdab," in Lexikon der Ägyptologie IV, 874-879.

_____. "Ahanakht of Bersheh and the Hare Nome in the First Intermediate Period and Middle Kingdom," in Studies in Ancient Egypt, the Aegean, and the Sudan: Essays in honor of Dows Dunham on the Occasion of his 90th Birthday. Boston: 1981, 14-30.

_____, Susan K. Doll, and Rita E. Freed. Egypt's Golden Age: The Art of Living in the New Kingdom. Boston: 1982.

Brown, J. Carter, et al. Treasures of Tutankhamun. New York: 1976.

Brunner, Hellmut. Die Texte aus den Gräbern der Herakleopolitenzeit von Siut. (Ägyptologische Forschungen 5.) Glückstadt-Hamburg-New York: 1937.

_____. "Maᶜa-cheru," in Lexikon der Ägyptologie III, 1107-1110.

Brunner-Traut, Emma. Der Tanz im Alten Ägypten. (Ägyptologische Forschungen 6.) Glückstadt: 1938.

_____. "Atum als Bogenschütze," in Mitteilungen des deutschen archäologischen Instituts Kairo 14 (1956), 20-28.

_____. "Esel," in Lexikon der Ägyptologie II, 27-30.

_____. "Gazelle," in Lexikon der Ägyptologie II, 426-427.

_____. "Gesten," in Lexikon der Ägyptologie II, 573-585.

_____. "Hase," in Lexikon der Ägyptologie II, 1023-1024.

_____. "Maus," in Lexikon der Ägyptologie III, 1250-1252.

_____. "Lotos," in Lexikon der Ägyptologie III, 1091-1096.

_____. "Spitzmaus," in Lexikon der Ägyptologie V, 1160-1161.

_____. "Tanz," in Lexikon der Ägyptologie VI, 215-231.

_____. "Aspective," in Heinrich Schäfer, Principles of Egyptian Art, translated from the German by J. Baines. Oxford: Griffith Institute; 1986.

Buchberger, Hannes. "Vogel," in Lexikon der Ägyptologie VI, 1046-1051.

Buhl, Marie-Louise. "The Goddesses of the Egyptian Tree Cult," in Journal of Near Eastern Studies 6 (1947), 80-97.

Butzer, Karl W. Early Hydraulic Civilization in Egypt. Chicago and London: 1976.

_____. "Wüste," in Lexikon der Ägyptologie VI, 1293.

_____. "Wüstentiere," in Lexikon der Ägyptologie VI, 1295-1297.

Caminos, Richard. Literary Fragments in the Hieratic Script. Oxford: 1956.

Curto, Silvio. "Krieg," in Lexikon der Ägyptologie III, 765-786.

D'Auria, Sue, Peter Lacovara, and Catharine H. Roehrig. Mummies and Magic: The Funerary Arts of Ancient Egypt. Boston: 1988.

David, A. Rosalie. A Guide to Religious Ritual at Abydos. Warminster: 1981.

Davies, Nina M. "Birds and Bats at Beni Hasan," in Journal of Egyptian Archaeology 35 (1949), 13-20.

_____, and Alan Gardiner. The Tomb of Amenemhēt. London: 1915.

Davies, Norman de Garis. "The Work of the Graphic branch of the expedition," in Bulletin of the Metropolitan Museum of Art 1933, April, section 2.

_____, and Alan H. Gardiner. The Tomb of Antefoker, Vizier of Sesostris I, and of His Wife, Senet. London: 1920.

Davis, Whitney. The Canonical Tradition in Ancient Egyptian Art. Cambridge: 1989.

Derchain, Philippe. Rites Égyptiens I: Le sacrifice de l'oryx. Brussels: 1962.

_____. "Snéferou et les Rameuses," in Revue d'Égyptologie 21 (1969), 19-25.

_____. "La Perruque et le cristal," in Studien zur altägyptischen Kultur 2 (1975), 55-74.

_____. "Symbols and metaphors in literature and representations of private life," in Royal Anthropological Institute Newsletter 15 (1976), 7-10.

Derchain-Urtel, Marie-Therèse. "Vögel, Aussenden der," in Lexikon der Ägyptologie VI, 1044-1046.

Doumas, F. "Fischer und Fischerei," in Lexikon der Ägyptologie II, 234-242.

Drenkhahn, Rosemarie. Der Handwerker und ihre Tätigkeiten im alten Ägypten. Wiesbaden: 1976.

_____. "Brot," in Lexikon der Ägyptologie I, 871.

_____. "Handwerker," in Lexikon der Ägyptologie II, 949-951.

Drioton, Étienne. "Une figuration cryptographique sur une stèle du Moyen Empire," in Revue d'Égyptologie 1 (1933), 203-229.

_____. Review of Grdseloff, in Annales du Service des Antiquités de l'Égypte 40 (1940/41), 1007-1014.

Dunham, D. "The Bird Trap," in Bulletin of the Boston Museum of Fine Arts 35 (1937), 52-54.

Eaton-Krauss, Marianne. The Representations of Statuary in Private Tombs of the Old Kingdom. Wiesbaden: 1984.

_____, and E. Graefe. The Small Golden Shrine from the Tomb of Tutankhamun. Oxford: 1985.

Edel, Elmar. Zu den Inschriften auf den Jahreszeitenreliefs der "Weltkammer" aus dem Sonnenheiligtum des Niuserre, Volumes I and II. Göttingen: 1961 and 1964.

Edwards, I.E.S. The Pyramids of Egypt. Harmondsworth: 1985.

Eggebrecht, Arne. "Gänseopfer," in Lexikon der Ägyptologie II, 371-372.

_____. "Schlachten;" "Schlachthof," in Lexikon der Ägyptologie V, 638-641.

Eggebrecht, Eva. "Greif," in Lexikon der Ägyptologie II, 895.

Emery, W.B. Archaic Egypt. Harmondsworth: 1961.

_____. A Funerary Repast in an Egyptian Tomb of the Archaic Period. Leiden: 1962.

Erman, Adolf. Life in ancient Egypt, translated from the German by H.M. Tirard. London: 1841.

_____. Reden, Rufe, und Lieder auf Gräberbildern des Alten Reiches. Berlin: 1919.

_____, and H. Grapow. Wörterbuch der Ägyptischen Sprache, 7 volumes. Leipzig: 1926-1963.

Faulkner, R.O. "The Rebellion in the Hare Nome," Journal of Egyptian Archaeology 30 (1954), 61-63.

_____. A Concise Dictionary of Middle Egyptian. Oxford: 1972.

_____. The Ancient Egyptian Coffin Texts, 3 volumes. Warminster: 1973-1978.

Finnestad, Ragnhild. Image of the World and Symbol of the Creator. Wiesbaden: 1985.

Fischer, Henry George. "The Nubian Mercenaries of Gebelein during the First Intermediate Period," in Kush 9 (1961), 44-80.

_____. Varia. New York: 1976.

_____. The Orientation of Hieroglyphs. Part I: Reversals. New York: 1977.

_____. "Bett," in Lexikon der Ägyptologie I, 767-768.

_____. "Fächer und Wedel," in Lexikon der Ägyptologie II, 81-85.

_____. "Gaufürst," in Lexikon der Ägyptologie II, 408-417.

_____. "Geißel," in Lexikon der Ägyptologie II, 516-517.

_____. "Hunde," in Lexikon der Ägyptologie III, 77-81.

_____. "Möbel," in Lexikon der Ägyptologie IV, 180-189.

_____. Egyptian Titles of the Middle Kingdom: A Supplement to Wm. Ward's Index. New York: 1985.

Frandsen, Paul John. "*Bwt* - Divine Kingship and Grammar," in BSAK 3 (1988), 152-158.

Franke, Detlef. "Probleme der Arbeit mit altägyptischen Titeln des Mittleren Reiches," in Göttinger Miszellen 83 (1984), 103-121.

Frankfort, Henri. Kingship and the Gods. Chicago: 1948.

_____. Ancient Egyptian Religion. New York: 1948.

Gaballa, G.A. Narrative in Egyptian Art. Mainz am Rhein: 1976.

Gaillard, C. "Identification de l'oiseau AMA figuré dans une tombe de Béni Hassan." Bulletin de l'Institut français d'Archéologie Orientale 33 (1933), 169-190.

Gamer-Wallert, Ingrid (Ingrid Wallert). Die Palmen in alten Ägypten (Münchner Ägyptologische Studien 1.) Berlin: 1962.

_____. "Baum, heiliger," in Lexikon der Ägyptologie I, 655-660.

_____. "Fische, religiös," in Lexikon der Ägyptologie II, 228-234.

_____. "Fishe, profan," in Lexikon der Ägyptologie II, 224-228.

_____. Fische und Fischkult im alten Ägypten. Wiesbaden: 1970.

Gardiner, Alan. The Attitude of the ancient Egyptians to Death and the Dead. Cambridge: 1935.

_____. "A Grim Metaphor," in Journal of Egyptian Archaeology 37 (1951), 29-31.

_____. Ancient Egyptian Onomastica, Volume II. Oxford: 1947.

_____. Egyptian Grammar, third edition. London: 1957.

_____. Egypt of the Pharaohs. Oxford: 1961.

Garstang, John. The Burial Customs of Ancient Egypt. London: 1907.

Germer, Renate. "Weihrauch," in Lexikon der Ägyptologie VI, 1167-1168.

Ghoneim, Wafik. Die ökonomische Bedeutung des Rindes im alten Ägypten. Bonn: 1977.

Goedicke, Hans. "Probleme der Herakleopolitenzeit." Mitteilungen des deutschen archäologischen Instituts Kairo 24 (1969), 136-143.

_____. Re-used Blocks from the Pyramid of Amenemhēt I at Lisht. New York: 1972.

_____. "Abi-Sha(i)'s Representation in Beni Hasan," in Journal of the American Research Center in Egypt 21 (1984), 203-210.

Grapow, H. Wie die alten Ägypter sich anredeten, wie sie sich grüssten und wie sie miteinander sprachen. Berlin: 1939-1943.

Grdseloff, B. Das Ägyptische Reinigungszelt. Cairo: 1951.

Grieshammer, Reinhard. Das Jenseitsgericht in den Sargtexten. Wiesbaden: 1970.

_____. "Mundöffnung(sritual)," Lexikon der Ägyptologie IV, 223-224.

Griffiths, J. Gwyn. The Origins of Osiris. Berlin: 1966.

_____. The Conflict of Horus and Seth. Liverpool: 1960.

Grimal, N.-C. La Stèle triomphale de Pi(ʿankh)y au Musée du Caire. (Études sur la propagande royale égyptienne, vol. I.) Cairo: 1981.

Guest, E.M. "Women's Titles in the Middle Kingdom. Ancient Egypt. (1926), 46-50.

Guglielmi, Waltraud. Reden, Rufe und Lieder...auf altägyptischen Darstellungen der Landwirtschaft, Viehzucht, des Fisch- und Vogelfangs vom Mittleren Reich bis zur Spätzeit. Bonn: 1973.

_____. "Ernte," Lexikon der Ägyptologie I, 1271-1272.

_____. "Fischerstechen," in Lexikon der Ägyptologie II, 242-244.

Gundlach, R. "Tempelrelief," in Lexikon der Ägyptologie VI, 407-411.

Habachi, Labib. Elephantine IV: The Sanctuary of Heqaib. Mainz am Rhein: 1985.

Haeny, G. "Zu den Platten mit Opfertischszene aus Heluan und Giseh," in Beiträge zur Ägyptischen Bauforschung und Altertumskunde 12 (Festschrift Ricke). Wiesbaden: 1971.

____. "Scheintür," in Lexikon der Ägyptologie V, 563-574.

Hagemann, Sabine. "Webstuhl," in Lexikon der Ägyptologie VI, 1160-1161.

Handoussa, Tohfa. "Fish offering in the Old Kingdom," in Mitteilungen des deutschen archäologischen Instituts Kairo 44 (1988), 105-109.

Harpur, Yvonne. Decoration in Egyptian Tombs of the Old Kingdom. London and New York: 1987.

Harris, J.R. Egyptian Art. London: 1966.

Hassan, Selim. Giza, 9 volumes. Cairo: 1932-1960.

_____. "Excavations at Saqqara 1937-38," in Annales du Service des Antiquités de l'Égypte 38 (1938), 503-521.

_____. "The Causeway of *Wnis* at Sakkara," in Zeitschrift für ägyptische Sprache und Altertumskunde 80 (1955), 136-137.

Hayes, William C. "The Middle Kingdom in Egypt," in The Cambridge Ancient History, Volume I, Part 2. Cambridge: 1971.

Heerma van Voss, Mathieu. "Sargtexte," in Lexikon der Ägyptologie V, 468-471.

Helck, H.W. Zur Verwaltung des Mittleren und Neuen Reiches. Leiden and Cologne: 1958.

_____. "Akazie," in Lexikon der Ägyptologie I, 113.

_____. "Bier," in Lexikon der Ägyptologie I, pp. 789-792.

_____. "Gaue," in Lexikon der Ägyptologie II, 385-408.

_____. "Gauzeichen," in Lexikon der Ägyptologie II, 422-426.

_____. "Opferstiftung," in Lexikon der Ägyptologie IV, 590-594.

_____. "Papyrusraufen, Papyruszepter," in Lexikon der Ägyptologie IV, 671-672.

_____. "Rituale," in Lexikon der Ägyptologie V, 271-285.

_____. "Weidewirtschaft," in Lexikon der Ägyptologie VI, 1166-1167.

_____, E. Otto and W. Westendorf (eds.). Lexikon der Ägyptologie, 6 volumes. Wiesbaden: 1972-1986.

_____. "Schlangenhalspanther," in Lexikon der Ägyptologie V, 52.

_____. "Zaubermesser," in Lexikon der Ägyptologie VI, 1355.

Hermann, A. Altägyptische Liebesdichtung. Wiesbaden: 1959.

Hickmann, Hans. Musicologie pharaonique. Baden: 1956.

_____. Musikgeschichte in Bildern: Ägypten. Leipzig: 1961.

Hölscher, Uvo. The Excavations of Medinet Habu, Volume III. Chicago: 1941.

Hölzl, Christian. Studien zur Entwicklung der Felsengräber: Datierung und lokale Entwicklung der Felsengräber des Mittleren Reiches in Mittelägypten. Dissertation, Wien: 1984.

Hornung, Erik. Conceptions of God in Ancient Egypt, translated from the German by J. Baines. Ithaca: 1982.

_____. The Valley of the Kings: Horizon of Eternity, translated from the German by David Warburton. New York: 1990.

_____. Idea into Image: Essays on Ancient Egyptian Thought, translated from the German by Elizabeth Bredeck. Princeton: 1992.

Houlihan, P. The Birds of Ancient Egypt. Cairo: 1986.

_____. "Some Remarks on Deer (Cervidae) in Ancient Egypt," in Journal of Egyptian Archaeology 73 (1987), 238-243.

Houtart, Albert. "Les Chiens dans l'Ancienne Égypte," in Chronique d'Égypte 17 (1934), 28-34.

Hugonot, Jean-Claude. Le jardin dans l'Egypte ancienne. Frankfurt am Main: 1989.

Iversen, Erik. Canon and Proportions in Egyptian Art. Warminster: 1975.

Janssen, J.J. Commodity Prices from the Ramessid Period. Leiden: 1975.

Jaroš-Deckert, B. Grabung im Asasif 1963-1970: Der Wandmalereien der XI. Dynastie. Mainz am Rhein: 1984.

Jéquier, Gustave. Les Frises d'objets des sarcophages du Moyen Empire. Cairo: 1921.

_____. Le monument funéraire de Pepi II, vols. I-III. Cairo: 1938.

Johl, C.H. Altägyptische Webestühle und Brettchenweberei in Altägypten. Leipzig: 1924.

Jones, Dilwyn. A Glossary of Ancient Egyptian Nautical Titles and Terms. London: 1988.

Junker, Hermann. Giza, 8 volumes. Vienna and Leipzig: 1929-1947.

Kaiser, W., et al. "Stadt und Tempel von Elephantine; 15./16. Grabungsbericht," in Mitteilungen des deutschen archäologischen Instituts Kairo 44 (1988), 135-182.

Kanawati, Naguib. "The Living and the dead in Old Kingdom tomb scenes," in Studien zur altägyptischen Kultur 9 (1981), 213-225.

Kaplony, Peter. "Ka," in Lexikon der Ägyptologie III, 275-282.

_____. "Toter am Opfertisch," in Lexikon der Ägyptologie VI, 713-726.

Kees, H. Totenglauben und Jenseitsvorstellungen der alten Ägypter. Leipzig: 1926.

_____. Ancient Egypt: A Geographical History of the New Kingdom. Chicago: 1961.

Kemp, Barry J. "Abydos and the royal tombs of the first Dynasty," in Journal of Egyptian Archaeology 52 (1966), 13-22.

_____. "The Egyptian 1st Dynasty Royal Cemetery," in Antiquity 41 (1967), 22-32.

_____. "Old Kingdom, Middle Kingdom and Second Intermediate Period c. 2686-1552 BC," in Ancient Egypt: A Social History. Cambridge: 1983.

_____. Ancient Egypt: Anatomy of a Civilization. London and New York: 1989.

Kessler, Dieter. Historische Topographie der Region zwischen Mallawi und Samalut. Wiesbaden: 1981.

_____. "Neferusi," in Lexikon der Ägyptologie IV, 383-385.

_____. "Menat-Chufu," in Lexikon der Ägyptologie IV, 41-42.

_____. "Her-wer," in Lexikon der Ägyptologie II, 1157-1158.

_____. "Zu den Jagdszenen auf dem kleinen goldenen Tutankhamun-schrein," in Göttinger Miszellen 90 (1986), 35-43.

_____. "Zur Bedeutung der Szenen des täglichen Lebens in den Privatgräbern (I): Die Szenen des Schiffsbaues und der Schiffahrt," in Zeitschrift für ägyptische Sprache und Altertumskunde 114 (1987), 59-88.

_____. "Die Asiatenkarawane von Beni Hassan," in Studien zur altägyptischen Kultur 14 (1987), 147-165.

_____. "Zur Bedeutung der Szenen des täglichen Lebens in den Privatgräbern (II): Schreiber und Schreiberstatue in den Gräbern des AR," in Zeitschrift für ägyptische Sprache und Altertumskunde 117 (1990), 21-43.

Klebs, Luise. Die Reliefs des alten Reiches. Heidelberg: 1915.

_____. Die Reliefs und Malereien des mittleren Reiches. Heidelberg: 1922.

_____. Die Reliefs und Malereien des neuen Reiches. Heidelberg: 1934.

Kuhlmann, Klaus P. Der Thron im alten Ägypten. Glückstadt: 1977.

Lapp, Gunther. Die Opferformel des alten Reiches. Mainz am Rhein: 1986.

Leclant, Jean. "Earu-Gefilde," in Lexikon der Ägyptologie I, 1156-1160.

_____, and A. Heyler. "Sur l'administration de l'Égypte au Moyen- et Nouvel-Empire," in Orient. Literaturz. 56 (1961), 118-129.

Lesko, Leonard H. "Ancient Egyptian Cosmogonies and Cosmology," in Religion in Ancient Egypt, Byron E. Shafer, ed. Ithaca: 1991.

Lichtheim, Miriam. Ancient Egyptian Literature, volumes I and II. Los Angeles: 1973 and 1976.

_____. Ancient Egyptian Autobiographies Chiefly of the Middle Kingdom. Göttingen: 1988.

Lloyd, Alan B. "Psychology and Society in the Ancient Egyptian Cult of the Dead," in Religion and Philosophy in Ancient Egypt, (Yale Egyptological Studies 3), W.K. Simpson, ed. New Haven: 1989.

Lucas, A. and J.R. Harris. Ancient Egyptian Materials and Industries, fourth edition. London: 1989.

Lüddeckens, Erich. "Untersuchungen über religiösen Gehalt, Sprache und Form der Ägyptischen Totenklagen," in Mitteilungen des deutschen archäologischen Instituts Kairo 11 (1943).

Manniche, Lise. Ancient Egyptian Musical Instruments. (Münchner Ägyptologische Studien 34.) Berlin: 1975.

Martin, Karl. "Vogelfang," in Lexikon der Ägyptologie VI, 1051-1054.

Martin-Pardey, E. "Schiff," in Lexikon der Ägyptologie V, 601-610.

_____. "Schiffbau," in Lexikon der Ägyptologie V, 616-619.

Meyer, Christine. "Wein," in Lexikon der Ägyptologie VI, 1169-1182.

Meyers, Elizabeth. A Program of Political Theology in Amarna Tomb Art. Ph.D. dissertation, University of Pennsylvania: 1981.

Michalowski, Kazimierz. Art of Ancient Egypt. New York: 1969.

Montet, Pierre. Les scènes de la vie privée dans les tombeaux égyptiens de l'ancien empire. Paris: 1925.

_____. Everyday Life in Egypt in the Days of Ramesses the Great. Philadelphia: 1988.

Mostafa, Doha M. "Reflexions sur la function cosmographique du temple égyptien," in Varia Aegyptiaca 5 (1989), 103-118.

Moussa, Ahmed and Hartwig Altenmüller. The Tomb of Nefer and Ka-hay. Mainz am Rhein: 1971.

Müller, D. "Der gute Hirte. Ein Beitrag zur Geschichte ägyptischer Bildrede," Zeitschrift für ägyptische Sprache und Altertumskunde 85 (1961), 126-144.

Müller, H.W. Die Felsengräber von Elephantine. (Ägyptologische Forschungen 9.) Glückstadt: 1940.

Müller, Maja. "Musterbuch," in Lexikon der Ägyptologie IV, 244-246.

Munro, Peter. "Größenverhältnisse," in Lexikon der Ägyptologie II, 903-906.

Naumann, R. "Der Tempel des Mittleren Reiches in Medinet Madi," in Mitteilungen des deutschen archäologischen Instituts Kairo 8 (1939), 185-189.

Naville, M., Percy Newberry, and George Willoughby Fraser. The Season's Work at Ahnas and Beni Hassan. London: 1891.

Needler, W. "Four relief-sculptures from the pyramid of Sesostris I at Lisht," in Annual (Art and Archaeology Division), Royal Ontario Museum, 1961, 15-26.

Nelson, Harold H. "The Egyptian Temple," in The Biblical Archaeologist 7 (1944), 44-53.

Newberry, P.E. "The Pig and the cult animal of Set." Journal of Egyptian Archaeology 14 (1928), 211-225.

_____, and others. Beni Hasan. Four parts (Archaeological Survey of Egypt, 1st, 2nd, 5th and 7th Memoirs). London: 1893-1900.

_____, and Griffith, F.Ll. El Bersheh. Two parts (Archaeological Survey of Egypt, 3rd and 4th Memoirs). London: 1893-1894.

O'Connor, David. "New Kingdom and Third Intermediate Period," in Ancient Egypt: A Social History. Cambridge: 1983.

_____. "The "Cenotaphs" of the Middle Kingdom at Abydos," in Mélanges Gamal Eddin Mokhtar. Cairo: 1985.

_____. "New Funerary Enclosures *(Talbezirke)* of the Early Dynastic Period at Abydos," in Journal of the American Research Center in Egypt 26 (1989), 51-86.

_____. "City and Palace in New Kingdom Egypt," in Cahier de recherches de l'Institut de Papyrologie et d'Egyptologie de Lille; Societies urbaines en Egypte et du Soudan 11 (1989), 73-87.

_____. "Nebhepetre's Chapel at Dendereh," paper given at ARCE, Seattle, 1992.

_____. "The Status of Early Egyptian Temples: An Alternative Theory," in The Followers of Horus: Studies dedicated to Michael Allen Hoffman, R. Friedman and B. Adams, eds., (Oxford and Bloomington, IN: 1992), 83-97.

_____. "Mirror of the Cosmos: the palace of Merenptah," in Proceedings of the Memphis Symposium on Ramesses II, R. Freed and E. Bleiberg, eds. (Memphis: 1993), 167-198.

Otto, Eberhard. "An Ancient Egyptian Hunting Ritual," in Journal of Near Eastern Studies 9 (1950), 164-177.

_____. "Abu Simbel," in Lexikon der Ägyptologie I, 25-67.

Parker, R.A. Ancient Egyptian Calendars. (Studies in Ancient Oriental Civilisation 26.) Chicago: 1950.

Piankoff, Alexandre. The Pyramid of Unas. Princeton: 1968.

Porter, Bertha, and Rosalind L. B. Moss. Topographical Bibliography of ancient Egyptian hieroglyphic texts, reliefs and paintings, 1st edition, volumes IV and VII. Oxford: 1934 and 1951; 2nd edition, volume II, Oxford: 1972.

Posener, Georges. Littérature et politique dans l'Égypte de la XIIe dynastie. Paris: 1956.

Posener-Kriéger, Paule. Les archives du temple funéraire de Néferirkari-Kakaï II. BdE 65.2, 1976.

Quibell, J.E. Excavations at Saqqara: The Tomb of Hesy. Cairo: 1913.

_____, and F.W. Green. Hierakonpolis II. London: 1902.

Quirke, Stephen. The Administration of Egypt in the Late Middle Kingdom. New Malden: 1990.

Ranke, Hermann. Die ägyptischen Personennamen, volumes 1-3. Glückstadt: 1935-1952.

Redford, Donald B. "The Father of Khnumhotpe II of Beni Hasan," in Journal of Egyptian Archaeology 53 (1967), 158-159.

Reeves, Nicholas and J.H. Taylor. Howard Carter: Before Tutankhamun. New York: 1993.

Reisner, George A. "The tomb of Hepzefa, nomarch of Siût," in Journal of Egyptian Archaeology 5 (1918), 79-98.

_____. The Development of the Egyptian Tomb Down to the Accession of Cheops. Cambridge, MA: 1936.

_____. A History of the Giza Necropolis. Cambridge, MA: 1942.

Reymond, E.A.E. The Mythical Origin of the Egyptian Temple. Manchester: 1969.

Ricke, H. and S. Schott. Bemerkungen zur ägyptischen Baukunst des Alten Reichs, in Beiträge zur ägyptischen Bauforschung und Altertumskunde, vol. 5. Cairo: 1950.

Roik, Eloise. Das ägyptische Wohnhaus. Frankfort am Main: 1988.

Rössler-Köhler, U. "Löwe," in Lexikon der Ägyptologie III, 1080-1090.

Roth, Ann Macy. "The Social Aspects of Death," in Mummies and Magic (Boston: 1988), 52-59.

Roth, H. Ling. Ancient Egyptian and Greek Looms. Halifax: 1951.

Rundle Clark, R.T. Myth and Symbol in Ancient Egypt. London: 1959.

Saad, Zaki Y. The Excavations at Helwan. Norman: 1969.

Säve-Söderbergh, Torgny. On Egyptian Representations of Hippopotamus Hunting as a Religious Motive. Uppsala: 1953.

_____. "Harpunieren," in Lexikon der Ägyptologie II, 1014-1015.

Schäfer, Heinrich. Principles of Egyptian Art, translated from the German by J. Baines. Oxford: 1986.

Scharff, Alexander. "Das Grab als Wohnhaus in der ägyptischen Frühzeit," in Sitzungberichte der Bayerische Akademie der Wissenschaft, 1944-1946, volume 6. Munich: 1946.

Scheel, Bernd. Egyptian Metalworking and Tools. Aylesbury: 1989.

Schenkel, Wolfgang. Frühmittelägyptische Studien. Bonn: 1962.

_____. Memphis Herakleopolis Theben; die epigraphischen Zeugnisse der 7.-11. Dynastie Ägyptens. (Ägyptologische Abhandlungen 12.) Wiesbaden: 1965.

_____. "Horus," Lexikon der Ägyptologie III, 14-25

Schmid, H.H. Gerechtigkeit als Weltordnung. Tübingen: 1968.

Schott, S. "Totenbuchspruch 175 in einem Ritual zur Vernichtung von Feinden," in Mitteilungen des deutschen archäologischen Instituts Kairo 14 (1956), 183-189.

Schulman, A. "The Battle Scenes of the Middle Kingdom," in Journal of the SSEA 12, no. 4 (1982), 165-183.

_____. "Siege Warfare in Pharaonic Egypt," in Natural History 73, nr. 3 (March 1964), 12-21.

Schweitzer, U. Das Wesen des Ka im Diesseits und Jenseits der alten Ägypter. Ägyptologische Forschungen 19 (1956).

Scott, Nora E. The Home Life of the Ancient Egyptians. New York: 1944.

Sethe, Kurt. Dramatische Texte zu altägyptischen Mysterienspielen. Leipzig: 1928.

Settgast, J. "Zu ungewöhnlichen Darstellungen von Bogenschützen," in Mitteilungen des deutschen archäologischen Instituts Kairo 25 (1960), 136-138.

_____. Untersuchungen zu altägyptischen Bestattungsdarstellungen. Glückstadt: 1963.

Shaheen, A. Historical Significance of Selected Scenes involving Western Asiatics and Nubians in the Private Theban tombs of the XVIIIth Dynasty. Ph.D. dissertation, University of Pennsylvania: 1988.

Shedid, Abdel Ghaffer. Die Felsgräber von Beni Hasan in Mittelägypten. Mainz am Rhein: 1994.

Silverman, David. "Textual Criticism in the Coffin Texts, in Religion and Philosophy in Ancient Egypt, (Yale Egyptological Studies 3), W.K. Simpson, ed. New Haven: 1989.

_____. "Divinity and Deities in Ancient Egypt," in Religion in Ancient Egypt, Byron E. Shafer, ed. Ithaca: 1991.

_____. "The Nature of Egyptian Kingship," in Ancient Egyptian Kingship, D. Silverman and D. O'Connor, eds. (Leiden, New York: 1995), 49-92.

Simpson, W.K. Terrace of the Great God at Abydos. New Haven and Philadelphia: 1974.

_____. "Kenotaph," in Lexikon der Ägyptologie III, 387-391.

Smith, E. Baldwin. Egyptian Architecture as Cultural Expression. New York and London: 1938.

Smith, H.S. "Animal Domestication and Animal Cult in Dynastic Egypt," in The Domestication and Exploitation of Plants and Animals. Peter J. Ucko et al., eds. Chicago: 1969, 307-316.

_____, and Alexandrina Smith. "A Reconsideration of the Kamose Texts," in Zeitschrift für ägyptische Sprache und Altertumskunde 103 (1976), 48-76.

Smith, William Stevenson. A History of Egyptian Sculpture and Painting in the Old Kingdom. Boston and London: 1946.

_____. "Paintings of the Egyptian Middle Kingdom at Bersheh," in American Journal of Archaeology 55 (1951), 321-332, pls. 18-21.

_____. The Art and Architecture of ancient Egypt. Harmondsworth: 1958.

Smither, B.P.C. "A Tax-Assessor's Journal of the Middle Kingdom," Journal of Egyptian Archaeology 27 (1941), 74-76.

Sorenson, Jørgen Podemann. "Divine Access: The so-called Democratization of Egyptian funerary literature as a socio-cultural process," in The Religion of the ancient Egyptians: Cognitive Structures and Popular Expressions, Gertie Englund, ed. Uppsala: 1989.

Spalinger, Anthony J. "A Redistributive Pattern at Assiut," in Journal of the American Oriental Society 105.1 (1985), 7-20.

_____. "Foods in P. Bulaq 18," in Studien zur altägyptischen Kultur 13 (1986), 207-247.

Spanel, Donald. "The Herakleopolitan Tombs of Kheti I, *Jt(.j)jb(.j)*, and Kheti II at Asyut," in Orientalia 58 (1979), 301-314.

_____. Beni Hasan in the Herakleopolitan Period. Ph.D dissertation, University of Toronto: 1984.

_____. "The Date of Ankhtifi of Mo'alla," in Göttinger Miszellen 78 (1984), 87-95.

_____. "Ancient Egyptian Boat Models of the Herakleopolitan Period and Eleventh Dynasty," in Studien zur altägyptischen Kultur 12 (1985), 243-253.

_____. Two Studies on the Herakleopolitan Period, forthcoming.

Spencer, A.J. Brick Architecture in Ancient Egypt. London: 1979.

_____. Death in Ancient Egypt. Aylesbury: 1982.

Spencer, Patricia. The Egyptian Temple: A Lexicographical Study. London: 1984.

Stadelmann, L. "Fledermaus," in Lexikon der Ägyptologie II, 263.

Staehelin, E. Untersuchungen zur ägyptischen Tracht im Alten Reich. (Münchner Ägyptologische Studien 8). Berlin: 1966.

_____. "Zur Bedeutung der Jagd im alten Ägypten," in Spiel und Sport im alten Ägypten. Basel: 1978.

_____. "Bart," in Lexikon der Ägyptologie I, 627-628.

Steindorff, G. "Haus und Tempel," in Zeitschrift für ägyptische Sprache und Altertumskunde 34 (1896), 107-110.

Störk, L. "Antilope," in Lexikon der Ägyptologie I, 319-323.

_____. "Elephant," in Lexikon der Ägyptologie I, 1214-1216.

_____. "Gans," in Lexikon der Ägyptologie II, 373-375.

_____. "Ginsterkatze," in Lexikon der Ägyptologie II, 597-599.

_____. "Hyäne," in Lexikon der Ägyptologie III, 91-92.

_____. "Karakal," in Lexikon der Ägyptologie III, 327.

_____. "Leopard," in Lexikon der Ägyptologie III, 1006-1007.

_____. "Mähnenschaf," in Lexikon der Ägyptologie III, 1123-1124.

_____. "Pflug-Pflügen," in Lexikon der Ägyptologie IV, 1013-1015.

_____. "Rind, " in Lexikon der Ägyptologie V, 257-263.

_____. "Steinbock," in Lexikon der Ägyptologie V, 1276.

_____. "Stachelschwein," in Lexikon der Ägyptologie V, 1231-1233.

Strauß-Seeber, "Spindel, Spinnen," in Lexikon der Ägyptologie V, 1156.

Teichmann, F. Der Mensch und sein Tempel. Stuttgart: 1978.

Tooley, A.M.J. "Coffin of a dog from Beni Hasan," in Journal of Egyptian Archaeology 74 (1988), 207-211.

Touny, A.D. and Steffen Wenig. Sport in Ancient Egypt, translated by Joan Becker. Leipzig: 1969.

Troy, Lana. Patterns of Queenship in ancient Egyptian myth and history. Uppsala: 1986.

Valloggia, M. "Les vizirs de XIe et XIIe dynasties," in Bulletin de l'Institut français d'Archéologie Orientale 74 (1974), 123-134.

Vandier, J. Mo'alla. La tombe d'Ankhtifi et la tombe de Sébekhotep. Cairo: 1950.

_____. Manuel d'archéologie égyptienne. 6 volumes. Paris: 1952-1978.

te Velde, H. Seth, God of Confusion. Leiden: 1967.

_____. "Seth," in Lexikon der Ägyptologie V, 908-11.

_____. "Horus und Seth," in Lexikon der Ägyptologie III, 25-27.

Vittman, Günther. "Orientierung," in Lexikon der Ägyptologie IV, 607-609.

Verhoeven, Ursula. Grillen, Kochen, Backen im Alltag und im Ritual Altägyptens. Brussels: 1984.

_____. "Totenmahl," Lexikon der Ägyptologie VI, 677-679.

Ward, William A. Index of Egyptian Administrative and Religious Titles of the Middle Kingdom. Beirut: 1982.

_____. "The Case of Mrs. Tchat and Her Sons at Beni Hasan," in Göttinger Miszellen 71 (1984), 51-59.

Weeks, Kent R. "Art, Word, and the Egyptian World View," in Egyptology and the Social Sciences. Cairo: 1979.

Wente, Edward F. "Hathor at the Jubilee," in Studies in Honor of John A. Wilson. Chicago: 1969 (Studies in Ancient Oriental Civilisation 35), 83-91.

_____. "Funerary Beliefs of the Ancient Egyptians," in Expedition 1982 (Winter), volume 24.

Westendorf, W. "Sonnenlauf," in Lexikon der Ägyptologie V, 1100-1103.

_____, "Weltbild," in Lexikon der Ägyptologie VI, 1211-1213.

Wiebach, S. Die ägyptische Scheintür: morphologische Studien zur Entwicklung und Bedeutung der Hauptkultstelle in den Privat-Gräbern des alten Reiches. Hamburg: 1981.

Wildung, Dieter. "Feindsymbolik," in Lexikon der Ägyptologie II, 146-148.

_____. "Garten," in Lexikon der Ägyptologie II, 376-378.

Willems, H. O. "The Nomarchs of the Hare Nome and Early Middle Kingdom History," in Jaarbericht van het Vooraziatisch-Egyptisch Genootschap 28 (1983-84), 80-102.

_____. Chests of Life: A Study of Typology and Conceptual Development of Middle Kingdom Standard Class Coffins. Leiden: 1988.

Wilson, Hilary. Egyptian Food and Drink. Aylesbury: 1988.

_____. "A Recipe for offering loaves?" in Journal of Egyptian Archaeology 74 (1988), 214-217.

Wilson, John A. "Funeral Services of the Egyptian Old Kingdom," in Journal of Near Eastern Studies 3 (1944), 201-218.

_____. "The Royal Myth in Ancient Egypt," in Proceedings of the American Philosophical Society, 100 (1956), 439-442.

Winlock, H.E. The Slain Soldiers of Neb-hepet-re Mentu-hotpe. New York: 1945.

_____. The rise and fall of the Middle Kingdom in Thebes. New York: 1947.

_____. Models of Daily Life in Ancient Egypt from the Tomb of Meket-Re at Thebes. Cambridge, MA: 1955.

Wolf, W. Die Kunst Ägyptens: Gestalt und Geschichte. Stuttgart: 1957.

Wreszinski, Walther. "Bäckerei," in Zeitschrift für ägyptische Sprache und Altertumskunde 61 (1926), 1-15.

Yadin, Yigael. "Hyksos Fortification and the Battering-ram," in Bulletin of the American School of Oriental Research 137 (Feb. 1955) 23-32.

_____. The Art of Warfare in Biblical Lands, 2 volumes. Jerusalem: 1963.

Yoyotte, Jean. Les pèlerinages dans l'Égypte ancienne. Sources Orientales 3: 1960.

Žabkar, L.V. A Study of the Ba Concept in Ancient Egyptian Texts. Studies in Ancient Oriental Civilisation 34 (1968).

INDEX

—A—

—B—

—C—

—D—

—E—

—G—

—H—

—I—

—K—

—M—

—N—

—O—

—P—

—Q—

—R—

—S—

—T—

—U—

—Z—

Plate 1: West Wall of Chapel, Tomb 3

Plate 2: North Wall of Chapel, Tomb 3

Plate 3: East Wall of Chapel, Tomb 3

Plate 4: South Wall of Chapel, Tomb 3

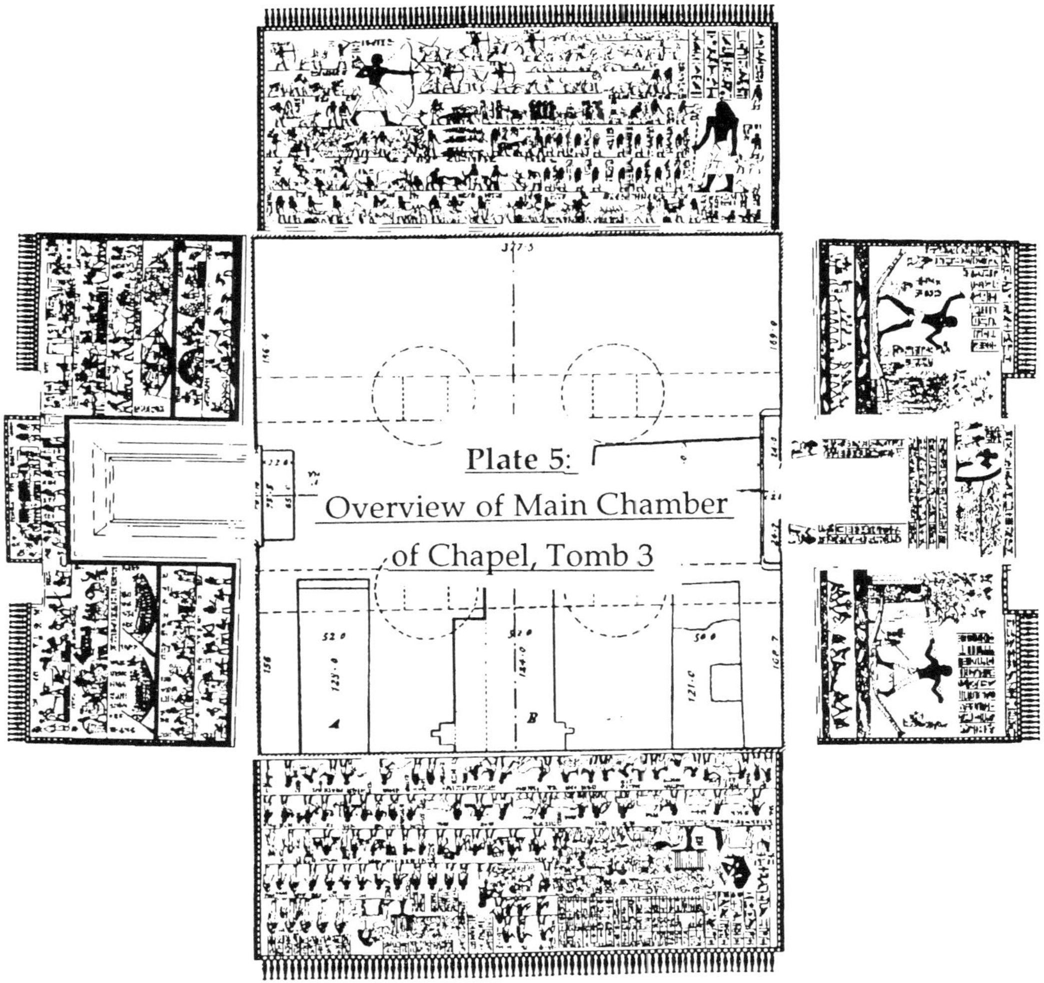

Plate 5:
Overview of Main Chamber of Chapel, Tomb 3

WEST WALL NORTH WALL

SOUTH WALL

Plate 6: Shrine of Chapel, Tomb 3